DANGEROUS RELATIONS

ADAM B. ULAM

Dangerous Relations

The Soviet Union in World Politics,
1970–1982

New York   Oxford
OXFORD UNIVERSITY PRESS
1983

Copyright © 1983 by Adam B. Ulam

Library of Congress Cataloging in Publication Data

Ulam, Adam Bruno, 1922–
  Dangerous relations.

  Includes index.
  1. Soviet Union—Foreign relations—1953–1975.
2. Soviet Union—Foreign relations—1975–
3. Detente. 4. World politics—1965–1975.
5. World politics—1975–1985. I. Title.
DK274.U4   327.47   82-14261
ISBN 0-19-503237-3

Printing: 9 8 7 6 5 4 3 2 1

Printed in the United States of America

Preface

Chronologically, this book takes up the story of Soviet foreign policy and of Soviet-American relations where they were left off in my *Expansion and Coexistence: Soviet Foreign Policy, 1917–73* (second edition, New York, 1974), and *The Rivals: America and Russia Since World War II* (New York, 1971), respectively. But insofar as its subject matter and approach are concerned, this is a self-contained study. It addresses the problem of detente and its vicissitudes and focuses on the dilemma of nuclear weapons and the revolution they have wrought in international relations. I have also sought answers to the questions of how the internal evolution of the Soviet system is likely to affect foreign policies of the USSR and what developments in the West might reverse the present unfortunate trends in world affairs, and best serve the interests of peace and international stability.

I must acknowledge with gratitude the help I received, while working on this book, from Christine Balm, Lubomyr Hajda, and Misha Tsypkin, my associates at the Russian Research Center at Harvard. And, as in the case of most of my previous books, this one owes a great deal to the congenial atmosphere of the Center, an institution unique in my experience, when it comes to combining intellectual stimulation with sociability.

Cambridge, Mass.   Adam B. Ulam
August 1982

Contents

DANGEROUS RELATIONS

1

Facts and Figures, 1945–70

Among the numerous government agencies that sprouted up in Washington just before Pearl Harbor, one stood out by its rather unusual name: the Office of Facts and Figures. There was nothing mysterious or clandestine about the agency. It had been created to provide a measure of official guidance to the news media, a virtual necessity with the country engaged in a global conflict, yet a government activity so much at variance with the American tradition in such matters that it was felt better to describe it in less than a forthright manner. With the war's end, the justification for any government management of news, be it ever so gentle and discreet, disappeared. The Office of War Information, which in the meantime had absorbed Facts and Figures, was dismantled.

There was something symbolic and ironic about the agency's name when taken in conjunction with the timing of its disappearance. It was precisely beginning in 1945 that Americans would need, more than ever before, reliable information about the outside world. In some ways the very expression "foreign" policy would become anachronistic, with what was happening in Berlin, Cuba, or Vietnam affecting the daily lives

3

of the people in this country as much, and almost as directly as, the events in their own backyard. Americans would have to strive to understand the unfamiliar world into which the US had been thrust by World War II and what followed it, which could hardly be called peace, as this term had been understood prior to 1939. But while groping for answers, both the policy-maker and the average citizen would too often overlook or ignore the facts and figures which might explain that world and instruct policies designed to cope with it.

The Soviet Union soon emerged as the most puzzling and perplexing force on the postwar international scene. The sudden shift of the Soviet colossus from the wartime alliance and collaboration to what seemed to be an attitude of unremitting hostility alarmed the US government and public opinion. And even after decades of experience in tortuous coexistence with the USSR, this sense of wonder and incomprehension of Kremlin policies and attitudes never quite disappeared on this side of the ocean. In January 1980, President Carter declared that he had learned more about the Russians within the few days after their invasion of Afghanistan than he had in his previous three years in office. This was the President of the United States, with his special sources of intelligence, innumerable expert advisers, and his own prior first-hand experience in dealing with Brezhnev and his colleagues! It is not that surprising that the average American's reactions to the Soviet phenomenon have not infrequently reminded one of the story of a visitor to a zoo who, contemplating a giraffe, shakes his head and says, "It is impossible."

While World War II was going on, most Americans approved of the actions of their own government and hence, by and large, of those of its allies. Occasional and jarring discord between US and Soviet leaders was, for the sake of preserving wartime morale, concealed from the public. There were those, in fact, who tended to believe at the time that peacetime America would likely have more trouble with "imperialist" Britain than with Communist Russia. Comradeship-at-arms, it was fondly believed, had dispelled the Kremlin's fears and suspicions about the capitalist West; it was clearly in Russia's interest to abide by the spirit and letter of the Charter of the United Nations.

Within a few months of VJ-Day, such facile optimism lay shattered. Tales of heroic deeds of Soviet soldiers and civilians were now succeeded by newspaper accounts of the Soviet depradations in Eastern Europe and of forcible imposition of Communism upon the nations that had barely emerged from the Nazi yoke, only to be subjected to a new one. As they tried to enjoy the boons of peace, and the rather unexpected prosperity it had brought, Americans felt unpleasantly constrained to turn their attention again to the international scene: their erstwhile ally was acting in a seemingly inexplicable, but increasingly ominous, fashion. Apart from Eastern Europe, the Soviet state appeared bent upon ideologico-military expansion elsewhere. In 1945 Moscow pressed territorial demands on Turkey; in 1946 refused in contravention of its wartime pledge, to pull its troops out of Northern Iran; and at the same time stirred up, through Bulgaria and its other then-docile satellite, Yugoslavia, the civil war in Greece. The USSR now remained the only great military power on the continent. All the more disquieting the fact (and it was held as such even by otherwise well-informed Western circles) that the Soviet army had been but partly demobilized and retained millions of men under arms when there was no conceivable threat to the security of Russia. And at the same time the French and Italian Communist parties, still exemplarily loyal to Moscow, were strong and bidding for power in their own right.

The initial puzzlement and irritation over the Soviets' actions soon gave way to alarm. The war years had not been conducive to a realistic appraisal of the Soviet system. Even for the most conservative American, the joint struggle dulled his instinctive anti-Communism. And now in the wake of shattered illusions, the reaction was all the more violent: Communism was seen as synonymous with tyranny. Stalin, the jovial and reasonable "Uncle Joe" of the wartime Big Three encounters, was now seen as a ruthless and fanatical tyrant. Extravagant hopes were succeeded by intemperate fears: that the ultimate objective of the Soviet policy-makers, who at the

same time were absolute masters of international Communism, was nothing short of a world state ruled by the Kremlin,
a goal in pursuit of which it was employing both military
power and ideological subversion.

The new situation confounded an American society whose
founders had foreseen only two types of relationship between
this country and foreign nations: "friends in peace, enemies
in war." The US and the USSR were not at war (in fact almost
uniquely among the great powers, America and Russia have
never been at war with each other); indeed, in the official diplomatic communications concerning Germany, etc., the Soviet
Union was still styled an "allied power." Yet very few Americans, even among those who put the main burden of responsibility for deteriorating relations with the Soviets on their own
country, by 1946 considered Stalin's Russia as America's
friend.

The term "Cold War" caught on in this period of disillusionment because it seemed the most convenient description
of the unusual relationship between what were now the only
two superpowers. Years later when many Americans chose to
vent their frustrations over the state of the world and American society on their own, rather than the Soviet, government
the term itself underwent a significant redefinition. No longer
was it assumed to describe the period of impasse in the relations of the two states, and the attempt of the US to cope with
the situation by all means short of a shooting war. The Cold
War, the proponents of the so-called revisionist school were
to assert, had little to do with America's or the free world's
security. It was fomented, so this theory went, by American
capitalism, which used the alleged Soviet-Communist danger
as a cover-up for its own predatory aims. "Cold warrior" became a term of opprobrium denoting anyone who would defend or advocate any of the policies the US adopted after the
war in order to contain the Soviet danger.

It is not proposed here to rehash once more the orthodox
versus revisionist debate about the Cold War.[1] Flourishing in

1. It is treated at length in my book *The Rivals* (New York, 1971; London,
1973), especially pp. 82–83, 93–97.

the era of Vietnam, the revisionist argument was more a political polemic than a dispassionate attempt to reassess a historical situation. The argument stands and falls by the validity of
its central thesis: it was the US government which, presumably at the behest and in the interest of what was subsequently
called the military-industrial complex, conjured up the nonexistent Soviet danger and then indoctrinated the American
public accordingly. Yet, as Arthur Schlesinger, at one time
sympathetic to the revisionist view, pointed out recently (cit-
ing approvingly a contemporary observation by a British diplomat): "[It was] the mass of ordinary people who first became
angry over Soviet actions and then turned the Truman administration around. . . . The driving force . . . has come not from
the top but from below. Events and public opinion have forced
the obviously uncertain and reluctant administration into af-
fording to the world at least some measure of the leadership
which the United States ought to be providing." [2]

We are thus drawn to a somewhat melancholy reflection:
powerful gusts of popular emotion rather than factual data
about the international situation determined the main lines of
American foreign policy. By 1947 most Americans had become convinced of the essential wickedness of Soviet Communism. As President Truman said in proclaiming the doctrine that bears his name, "[It is] based upon the will of a
minority forcibly imposed upon the majority. It relies upon
terror and oppression . . . and the suppression of personal
freedoms." [3]

Given then that Soviet Communism was not only totalitar-
ian but imperialistic, did it necessarily follow that the Soviet
Union was *already* a clear and present danger to the security
of this country? Not any more than the fact that the Russian
people had fought heroically against Hitler should have been
used as a proof that Stalin's government would be America's
reliable partner in peace; nor any more than the trauma of

2. Quoted in Arthur M. Schlesinger, "The Cold War Revisited," *New York
Review of Books,* October 25, 1979.

3. Harry S. Truman, *Memoirs,* vol. 2, *Years of Trial and Hope* (New York,
1958), p. 106.

Vietnam offered a prima facie proof that in the sixties and seventies it was US rather than Soviet policies that threatened the world order. In the late forties the US hugely overestimated Soviet military power, just as in the sixties it was to overlook, because of its domestic preoccupation, the prodigious increase in Russia's conventional and strategic arms. And as we enter the eighties there is an unfortunate possibility that American policies will become unduly attuned to the statistics of Soviet ICBMs and tanks, rather than considering them within the general context of facts and figures that underlies the precarious coexistence of the two superpowers.

It was a basic premise as well as a gnawing concern of the American policy-makers of the Cold War era that if the USSR were to be directly challenged, it might resort to war. Symptomatic of Washington's fears was the rather pathetic question Walter Bedell Smith, our ambassador to Moscow, asked Stalin in April 1946: "What does the Soviet Union want and how far is Russia going to go?"[4] "Not much further," replied the dictator, undoubtedly secretly amused and reassured. Even Gen. Lucius Clay, the US commander in Germany, usually prone to call a Soviet bluff, could become agitated to the point of wiring Washington on March 5, 1948, that "war may come with dramatic suddenness."[5] When the Russians did cut off land access to West Berlin, Washington rejected any suggestions (including General Clay's) that it should challenge the blockade directly, and resorted instead to an airlift to keep the city supplied with food and fuel. The Soviet Union, it was felt, must not be confronted with anything like an ultimatum, even in the face of such direct provocation.

Such fears were based upon a conviction that the Kremlin was both ready and willing to risk a military confrontation with the Western powers. Though they must have had some realistic intelligence on the subject, Western policy-makers could not free their minds of awe of the Soviet military might, nor of the vision of millions of Soviet soldiers overwhelming

4. Walter Bedell Smith, *My Three Years in Moscow* (New York, 1950), p. 47.

5. W. Philips Davison, *The Berlin Blockade* (Princeton, 1958), p. 73.

the feeble American and British contingents and sweeping rapidly through their zones in Germany, indeed all the way to the English Channel. It was widely believed, even by the people who were in a position to know better, that only America's monopoly of the bomb restrained the Kremlin from outright aggression, and even so the as-yet-modest US stockpile of nuclear weapons was thought to offset only partially the crushing Soviet military superiority.

It should not have required special intelligence sources to recognize that such views of Russian intentions and capabilities verged on the preposterous. In 1948 Nicholas Voznesensky, deputy prime minister of the USSR and its chief economic administrator, published a work detailing what the war had done to his country. The book adduced some somber statistics. Voznesensky testified that in the course of the war "31,850 Soviet factories, mills and other industrial enterprises exclusive of small plants" were either completely or largely destroyed and sacked. Equally disastrous were the losses in agriculture: "1,876 state farms, 2,890 machine tractor stations, and 98,000 collective farms" were completely devastated. (The total number of collective farms prior to the war was about 200,000, and the ones stricken were, for the most part, in the most fertile regions of the USSR.)[6]

In Voznesensky's book there were also other figures of great interest to anyone pondering whether the USSR might risk war with the US: "[The dollar value of] American production of armaments increased from 32.5 billion in 1942 to 60 billion in 1943; manufacture of aircraft of all types expanded from 47,900 in 1942 to 85,900 in 1943; mercantile shipping advanced from 5.1 to 12.1 million tons, while the tonnage of warships launched increased from 859,000 in 1942 to 2.61 million tons in 1943."[7]

Here then on the one hand was the US, whose enormous production potential had been unleashed by the war and which was ready in an emergency to surpass its industrial effort of

6. N. A. Voznesensky, *The Soviet Economy During the Second World War* [English translation] (New York, 1949), p. 131.

7. *Ibid.*, p. 20.

1942–45, and on the other the Soviet Union, its economy crippled and desperately in need of a lengthy period of peace just to make up its wartime losses. That could not have been the intended moral of Voznesensky's book, and perhaps the fact that inadvertently he revealed Russian industrial (and hence military) weakness contributed to his subsequent doom. (First dismissed from all his offices, he was ordered shot in 1950, and his book was condemned for its "un-Marxist" approach.) But Stalin, probably more than any other contemporary national leader, tended to gauge military power by a country's indices of industrial production—so much so that he declared in 1946 that only when the USSR reached an annual output of 60 million tons of steel could it achieve a measure of military security. In 1945 the actual figures were 12.3 million tons for the Soviet Union, as against 75 million for the US.

What about those millions of men that the USSR allegedly kept under arms while America demobilized at a frantic pace? Common sense should have urged the utter implausibility of the Russians retaining their army at anything like its wartime level. To do so would have in fact perpetuated the country's desperate economic condition and threatened its political stability. At the time the full and appalling extent of Russia's human losses was not known in the West. But whether the actual figure was seven million, as the Soviets then gave out, or over twenty million, as they have since admitted, it should have been obvious that the USSR had every reason to demobilize almost as rapidly as the US. Young men and women were badly needed on the farms and in the factories. How then could the government of a country that had just lost over 10 percent of its prewar population even think of embarking on a new war? And that against a state whose manpower resources had barely been scratched and whose industrial production surpassed the USSR's own several times? And in fact we have no reason to question Khrushchev's 1960 statement that, while at the end of the war Russia's armed forces comprised 11,365,000 people (fewer, incidentally, than those at the time under arms in the US), they were "as a result of demobilization carried out immediately after the war" reduced by 1948 to 2,874,000.[8] In view of the extensive military needs to garrison East Germany and the satellites, and domestic security requirements, the latter figure ought to dispel any notion of the Kremlin preparing to launch or being capable of launching a rapid strike against Western Europe. At the same time, the low figure of Soviet troops under arms suggests that the Russian leadership was not unduly fearful of imperialist aggression.

The Berlin blockade, which began in June 1948, convinced many outside observers that the Russians were quite ready to go to war. In fact the blockade could have been lifted instantaneously had the US given the slightest indication that it would challenge it directly (say, by announcing it would send an armed convoy through East Germany). There were, of course, obvious risks for the USSR in interdicting the West's land route to Berlin, but Stalin believed, and it is impossible to say he was wrong, that he could gauge the mood and probable reactions of Washington and London. It is difficult for democratic leadership to countenance a course of action which involves even a small risk of precipitating armed hostilities, while a dictatorship such as the Soviet one can make a credible show of *appearing* unconcerned about what for a Western democracy would be appalling risks.

We shall never know, unless some miracle enables us to read the protocols of the Politburo discussions, to what extent Russian equanimity in the face of America's crushing military-industrial superiority was based on Stalin's intuitions as against solid data gleaned by Soviet intelligence. The latter was in a position to know much about Western decision-making through its secret agents such as the notorious Burgess, McLean, and Philby (three British officials who spied for the Russians). But it is unreasonable to see Soviet espionage as the main factor in Moscow's by and large successful game of bluff. No one reading the American and British press of the period (1945–48) or following contemporary Congressional and Parliamentary debates would have concluded that what the Soviets were doing in Eastern Europe (i.e., grinding down the

8. *Pravda*, January 15, 1960.

remnants of opposition to Communism in East Germany, Poland, Czechoslovakia) would be met by the West with anything more than diplomatic protests (and even those would not last very long). One of the more extravagant assertions of the revisionist school has been the claim that the US employed its monopoly of the bomb to pressure the Russians. Indeed some of these authors[9] saw the decision to drop the bomb on Hiroshima and Nagasaki as motivated by Washington's intention to make Moscow more tractable rather than by a desire to bring the war against Japan to a swift conclusion. Yet no one has presented a single piece of evidence showing that the US ever employed its then monopoly of nuclear weapons to wrest concessions from the USSR. And even more to the point, no one has explained what it was that the American atomic blackmail allegedly prevented the Kremlin from doing. To be sure, Americans, whether in the government or not, were considerably surprised at the Soviets' apparent nonchalance over US possession of the dreaded weapon, and felt that had the roles been reversed *they* certainly would have been scared and inclined to appease the other party. But as has already been argued, it was public opinion in the West which was frightened about what the Russians might do over and above what they were already doing rather than vice versa. In an interview he gave in 1946 to a Western newspaperman, Stalin showed considerable composure about the whole subject: "Atom bombs are designed to scare those with weak nerves, but they cannot decide wars because there are not enough of them. To be sure atomic monopoly is a threat, but against it are two remedies: (a) monopoly of the bomb will not last, (b) the use of the bomb will be banned."[10]

The doctrine of containment, which provided the main analytical prop for American policies during what might be called the classical period of the Cold War (1947–53), was thus basically defensive in character. It did not seek to force or persuade the Russians to change the pattern of their international

9. E.g., Gar Alperovitz in his *Atomic Diplomacy: Hiroshima and Potsdam* (New York, 1965).

10. Stalin, *Works* (Stanford, 1955), vol. 3, p. 52.

behavior, say in Eastern Europe, where the process of "satellization" of the individual states was in 1947–48 still some way from being completed. The doctrine urged the necessity of deterring the Soviet Union from *fresh* imperialist ventures. Even so, this deterrence was to be circumspect enough so as not to provoke the Kremlin. Containment, wrote George Kennan, the main architect of the doctrine, meant that the West ought "to confront the Russians with unalterable counter-force at every point where they show signs of encroaching upon the interests of a peaceful and stable world"; but the US's "demands on Russian policy should be put forward in such a manner as to leave the way open for a compliance not too detrimental to Russian prestige." [11] Kennan did not, however, propose that the West probe or exploit Soviet weaknesses and vulnerabilities. The United States must remain on the defensive, reacting only when the Kremlin attempted to breach the free world's defenses. Only after many such, hopefully unsuccessful, attempts could one expect the Soviets to desist: "For no mystical Messianic movement, and particularly not that of the Kremlin, can face frustration indefinitely without eventually adjusting itself in one way or another to the logic of that state of affairs." [12]

The doctrine of containment, then, preached the need for patience during what was assumed must be a prolonged siege of the free world by the USSR and Communism (and at the time the two were considered virtually synonymous). The Soviets would continue to probe for weak spots in the West's defenses without the West reciprocating vis-à-vis Russia. The free world had to put its own house in order, economically and militarily, so as to repel, and eventually discourage, Soviet assaults. The Kremlin must be left in no doubt that any military incursion beyond the borders of the Soviet zone in Germany would mean war; hence the founding of the North Atlantic Treaty Organization in 1949. Countries fighting Communist insurgency or subjected to Soviet pressure for territo-

11. "The Sources of Soviet Conduct" (published originally in July 1947), in George Kennan, *American Diplomacy, 1900–1950*, pp. 121, 107.

12. *Ibid.*, p. 124.

rial concessions must be assured of American help, and thus the Truman Doctrine extended this help to Greece and Turkey. The massive infusion of economic assistance to Western Europe was only partly motivated by Cold War exigencies. Still, by helping the recovery of the war-ravaged economies, the Marshall Plan, it was hoped, would undercut the appeal of Communism, especially in France and Italy.

In sum, an assortment of policies was designed to contain Soviet Communism. By the same token, none of them as perceived on this side of the ocean ought to have been interpreted as a threat to Soviet security or hegemony in Eastern Europe. In fact the USSR and two of its satellites were originally invited to participate in the Marshall Plan and thus, presumably, to be beneficiaries of American economic help.

In retrospect it is difficult to fault the *concept* of containment. Granted the nature of the American political system it was probably the only practical response to the challenge of the postwar international situation. In theory it is possible, of course, to sketch different scenarios of American-Soviet relations: the US confronting Moscow with an ultimatum that it must genuinely abide by its wartime pledges concerning Eastern Europe or else; or, conversely, an agreement on spheres of influence between the two superpowers, the USSR being given a free hand concerning Poland, Czechoslovakia, etc., but desisting from any designs to advance its power or to promote Communism elsewhere. But to sketch such alternatives is to see immediately their utter unreality. They might have been possible had the US been a dictatorship or a nineteenth-century style empire. But what was within American *power* was entirely unfeasible given the psychology and the institutional framework of American *democracy*. The American people wanted policies that promised to arrest Soviet expansion and the spread of Communism. What they emphatically did *not* want was to carry the burden of responsibility for any course of action that entailed even a small risk of war. Thus containment seemed an ideal compromise between the two imperatives.

By 1949 it became obvious that this carefully thought-out

policy was flawed: the world's most populous nation had become Communist. And one year later another basic premise of
America's new foreign policy was found wanting. The US was
at war, not with the USSR to be sure, but with its North Korean clients, and later on with its then close allies, the Chinese
Communists. Thus containment had failed both to contain
Communism and to keep the US out of a shooting war. Was it
the idea that was wrong, or did the trouble lie in its execu-
tion?

We must look briefly at the other, Soviet, side of the picture.
As has already been shown, far from being ready to set on a
course of conquering the world for Communism, the USSR
badly needed and wanted peace. Yet the Kremlin was astute
enough to perceive that were the US to realize how desperate
was that need and how consequently strong was its own position vis-à-vis Moscow, the latter would be subjected to
mounting American pressures. The only way to offset the
crushing American superiority was to act as if it did not exist,
to make it appear that yes, the Soviet Union was ready to go
to war rather than to relax its grip on Eastern Europe or to
abide there and elsewhere by its wartime pledges. Like the
policy of containment for America, so the posture of ominous
isolation was the most logical postwar policy for Russia. To
appear *entirely* reasonable and peaceful, to continue the wartime ties and amicable relations with the West, would have
put in jeopardy, in Stalin's judgment, not only his newly acquired East European empire, but perhaps the security of the
Soviet Union itself. The USSR did not see itself as yet strong
enough to be friendly with the democracies.

The most vivid demonstration of this was the Soviet reaction to the Marshall Plan. Superficially the Kremlin's final refusal (after some initial hesitations) to join in the European
recovery plan must have seemed the height of irrationality.
What did the USSR have to lose by expressing its willingness
to participate? If the plan was then approved by Congress,
Russia, along with Western Europe, would become a recipient
of American economic assistance, which might speed up its
own recovery. On the other hand, if Congress (the Republi-

cans then enjoyed a majority in both houses) became scandal-
ized at the prospect of American billions helping Communist
states, it might reject the plan as a whole. Western Europe's
economic resuscitation would then be delayed, something
clearly desirable from the Communist point of view.

But Stalin knew better. Even to express willingness to accept capitalist largesse would have damaged the image of Soviet power which he meant to project both abroad and at home.
It is unusual to stand in awe of somebody who is a recipient
of your bounty. And participation in the plan would have required the USSR to be forthcoming with detailed data about
its industrial and agricultural production, which would have
probably revealed that the Soviet government systematically
falsified its own economic statistics and thus bared to the West
the country's extreme economic weakness.

At home the resumption of close contacts with the Western
democracies (and how could you take American money without inviting visits by Congressional delegations, experts, etc.?)
would have clashed with the Kremlin's determination to reim-
pose the full rigor of ideological and intellectual controls that
had been somewhat relaxed amidst the patriotic exaltation of
the war period. Not even during the great terror and spy mania
of the thirties had the Soviet Union been as hermetically sealed
off from the outside world, whether in terms of personal, intellectual, or even scientific contacts, as it would be between
the war's end and Stalin's death in 1953. In one of Winston
Churchill's most penetrating insights he declared that the
Kremlin feared the West's friendship no less than its enmity.
Such friendship, or rather its consequence—the Soviet citizen's exposure to Western ideas and contacts—might, in his
rulers' opinion, have struck at the very foundations of the
Communist system.

The basic flaw of containment thus lay not in the concept
itself, but in the fact that policies developed under it did not
take into account the full extent of Russia's fears and weaknesses. Had he realized the strains under which the Kremlin
found itself because of its imperialist ventures, Kennan wrote
many years later, he "would have had a far stronger case for

challenging the permanency of the imposing and forbidding facade which Stalin's Russia presented to the outside world in those immediate post-war years." [13] He was referring mainly to the strains created by Moscow's "involvement in Eastern Europe," [14] and therein lies the tale of Russia's continuous vulnerability and of America's recurring and lost opportunities ever since the end of World War II.

Like all the great conflicts of the twentieth century, the Cold War was triggered, if not exactly caused, by developments in Eastern Europe. It is still generally believed that already during the war this area had been conceded to the USSR and that the American and British governments, or more precisely Roosevelt and Churchill, by what they both said and implied at Teheran and Yalta, left Stalin a free hand in regard to the whole area from Poland to Bulgaria. At best this is a huge exaggeration. As the war was drawing to its close Roosevelt and Churchill, like most of the realistic observers of the war scene, had to conclude that the USSR would be the dominant power in the area.[15] But the general expectation was that Russia's domination would not be absolute, that there would be established a sphere of influence in the nineteenth-century meaning of the term: Poland, Czechoslovakia, etc., having to defer to Moscow's wishes insofar as their foreign and defense policies were concerned, but not being compelled against the wishes of the great majority of the local population to adopt the Communist one-party model. It was the brutality with which the process of satellization was carried out by the Soviets and their local Communist agents that became crucial in producing the emotional climate of the Cold War.

Yet, as we have also seen, it was generally assumed in the West that nothing could be done about Eastern Europe; that a determined attempt to prevent the Kremlin from turning its unfortunate neighbors into its servile fiefdoms would have led to war.

13. George F. Kennan, *Memoirs, 1925–1950* (Boston, 1967), p. 358.

14. *Ibid.*

15. And Churchill did hope to save some British influence there. See my *Expansion and Coexistence,* 2d ed. (New York, 1973), pp. 364–66.

By 1947 satellization was still far from complete. Only one country had already been turned into a close copy of the USSR, and that, ironically, was Yugoslavia, where the main impetus for rapid elimination of all barriers to one-party dictatorship had come from the local Communists. Elsewhere, without the presence or shadow of the Red Army behind the Kremlin's local satraps, free institutions and a genuine multiparty system still could have been saved.

The conviction that short of war the Soviets would not be budged from any area the Red Army had come to occupy flew in the face of two recent examples of Soviet withdrawal or restraint under precisely such circumstances. There was, first, the case of Finland. Historically and strategically this small nation offered more reasons for the Kremlin to try to turn it into a satellite, if not indeed to incorporate it outright into the USSR, than, say, Hungary or Romania. Finland had been a part of Russia's domains between 1809 and 1918.[16] Stalin had sought to conquer it in 1939–40, and later Finland joined with Germany in the attack on the Soviet Union. The USSR had, and until 1955 would keep, military bases on the territory of its northern neighbor. Finland's own Communist Party, while not wildly popular, had certainly a bigger genuine following among the population than did those of Poland and Hungary. Yet for all such eminent qualifications to become a "people's democracy," as the satellites were being styled, or indeed even a union republic of the USSR, Finland was allowed to preserve internal freedom while remaining in the Soviet sphere of influence. In foreign policy and defense Finland had to heed its colossal neighbor, but no determined attempt was made to impose a Communist regime there, and even a subsequent ejection of the Communists from the coalition government (in 1948) did not bring a Soviet intervention. For a regime as gluttonous for power and territory as Stalin's, indeed a most unusual restraint! Why?

Very likely it was because of the United States. We might call this a case of unwitting containment. Most Americans

16. Even if in law it was only a grand duchy joined by personal union to the Russian empire.

might have forgotten, but Stalin had not, the wave of anti-Soviet indignation that swept the US following the Soviet attack upon "little Finland" in 1939, indignation much greater in its emotional intensity than that which a few months earlier had greeted the news of the Hitler-Stalin deal. Subsequently, even though Finland became Hitler's ally, the US never declared war upon it. Beginning in 1944, the forcible imposition of Communism on Eastern Europe had been carried out gradually and carefully, with the Kremlin watching American reactions and adjusting its tactics accordingly. Finland might have been left for the final phase of the process, but by then—1948—other events, first the Berlin blockade and then the Korean War, made an additional challenge to American susceptibility too risky.

The US was directly involved in another case of the Soviets' being constrained to abandon their apparent prey. A wartime agreement (signed, appropriately enough, in Teheran in 1943) pledged the Big Three to pull their troops out of Iran within six months after the end of the war. Britain and the US did so as scheduled. But the USSR, which had occupied the northern part of the country, showed no signs of following suit. In fact in December 1945, two "independent" regimes, the Autonomous Republic of Azerbaijan and the Kurdish People's Republic, were established in the north under Soviet sponsorship. Iran appeared destined to share the fate of such truncated countries as Germany and Korea, its northern part becoming a Soviet satellite. As in other areas, the USSR sought to reclaim the imperial heritage of prerevolutionary Russia, which had long sought, and in 1907 had obtained, a virtual protectorate over the northern tier of what was then called Persia. Iran's government's expostulations and the Western powers' diplomatic protests were having no more effect than similar representations about Soviet usurpations in Eastern Europe. But in the case of this oil-rich and strategically located country, American diplomacy was proving more tenacious than elsewhere, as indicated by the fact that the issue came before the Security Council of the UN on March 25, 1946. One day later the Soviet representative, Gromyko, promised that Russian troops would be pulled out within six weeks. Following the withdrawal, the two separatist regimes were easily suppressed by the Iranian forces and their leaders put to death. For the moment an attempt to repeat the Eastern European scenario in the Middle East was frustrated. The alleged axiom that the Soviets never surrendered an area in which they had established military presence had been disproved.

Yet no lessons were drawn from the Soviet discomfiture in Iran. Why did the Russians pull out and abandon certainly a more glittering prize than, say, Bulgaria, one contiguous to the Soviet Union, part of Russia's sphere of influence, and hence, according to Stalin's logic, a legitimate target of expansion? The reason could not have been the anticipated censure by the Security Council, which the USSR would have vetoed, thus depriving it of any concrete effect. But the Kremlin perceived a significant difference between the American attitude toward Russia's foray into Iran and that concerning the Soviets' tightening grip on Eastern Europe. The latter the American public was now ready to concede as being one of the facts of international life. In contrast, Iran was a "new" issue. Having aroused American public opinion about this new and flagrant breach of their treaty obligations, the Soviets risked reawakening opposition to their previous conquests. There was an additional reason for Moscow's anxiety and concern. On March 5 Winston Churchill had made his famous "iron curtain" speech in Fulton, Missouri, in which he called on the Western world to unite in the face of the Soviet threat. Churchill's message was received rather coldly among the liberal circles of the US and Great Britain. But another example of the Soviets' scorning their pledges might—in conjunction with the old warrior's speech—have tipped the precarious balance between apathy and moral indignation in the West's view of Soviet actions. The American colossus might have been made aware of its own strength and Russia's weakness. Accidentally and unwittingly, the Western powers stumbled upon a means of making the USSR pull back: speaking harshly and insistently. This was not a case of actual confrontation between the two powers, but rather a situation out of which such a confrontation might grow if the weaker power did not yield, and in this case the USSR did.

Its restraint over Finland and retreat on Iran were made easier for the Kremlin because of the fact that neither was seen by the West as indicating Russia's weakness. Yet in both cases it was obviously the fear of what the US *might* do which accounted for the Soviets' atypical moderation. The story of northern Iran in 1946 is especially suggestive of what tenacious, well-informed American diplomacy could have accomplished in stopping or moderating Soviet expansion. Yet containment, as we have seen, virtually dismissed diplomacy as a means of altering Soviet behavior. By 1947 it had become virtually an article of faith within the State Department that negotiating with the Russians led nowhere, or down a primrose path. And this seemed borne out by the actual record of negotiations since the war's end: they would quickly turn into angry altercations, with the subject under discussion, e.g., the German peace treaty, remaining unresolved—or, as in the case of the puppet regimes of Eastern Europe, the Western negotiators giving in to the Soviets out of sheer weariness. American diplomacy suffered from two great handicaps in dealing with the USSR: it failed to impress on the Kremlin that it was aware of its own enormous strength and of Russia's weakness; it also lacked patience.

Those initial sins of omission have set, disastrously, the tone of American-Soviet relations down to our own day. Few would claim that even the most astute and patient US foreign policies could have assured an entirely peaceful world or prevented occasional and sharp conflicts between the two superpowers. But it is not unreasonable to argue that had this country entered the postwar era with a more realistic appraisal of the Soviet Union, its goals, fears, and potentialities, that in itself would have saved the democratic world many of the dangers and discomfitures it has experienced over the last thirty-odd years. As this is being written (1982) Eastern Europe is once more a major cause of international tension, largely because of what the USSR did, and the US failed to do, concerning this area in the 1940s. Had the US, as was then within its power, restrained the Soviet Union from imposing Communism upon Poland, and that country, like Finland, been allowed to retain its internal freedom, the Kremlin would be spared its present dilemma. Should the present martial-law policy of the Warsaw regime fail, the Soviets might yet have to intervene militarily or tolerate the virtual emasculation of the local Communist regime, either choice involving incalculable risks and costs, and not merely for Poland and Russia.

Not being able to modify Soviet international behavior at the time of greatest US superiority has thus been the prime cause of this country's discomfitures in foreign policy. Having secured American acquiescence to its imperial expansion at the period of its most pronounced weakness, Moscow was not going to alter its policy when, as Russia began to recover from the war's ravages, it grew progressively stronger. One of the central premises behind containment was that it would eventually constrain the USSR to behave in a fashion consonant with its obligations under the UN Charter, for the Kremlin could not "face frustration indefinitely without adjusting itself in one way or another to the logic of that state of affairs." Yet there was no reason for Stalin to feel frustrated or chastened over a situation which had enabled him to achieve the goal that had eluded the empire of the tsars for two centuries: Russia's absolute hegemony over a huge area of East Central Europe, with its more than 100 million inhabitants. There could be no better advertisement, both at home and abroad, of the viability and dynamism of the Communist system than this brilliant achievement of Soviet diplomacy, hence no incentive for the Soviet leaders to devote most of their efforts and resources to the improvement of their subjects' lot, rather than to imperial expansion.

Ironically, the only major frustration Stalin's Russia experienced in foreign relations came in its dealings with a small Balkan state, Tito's Yugoslavia, rather than on account of any actions by the powerful United States. The outcome of the Berlin blockade, usually described as a defeat for the Kremlin, could more justifiably be seen as a draw: the USSR did not succeed in making the West desist from creating the German

Federal Republic, but it did, and with impunity, violate Western rights and its treaty obligations. Nor was it the Soviet
Union that was made to suffer the consequences of licensing
(if indeed not ordering) North Korea's invasion of the South.
Both acts of provocation, though they did not accomplish their
primary objectives, still furnished Soviet policy-makers with
useful lessons. Berlin was shown to be an exposed nerve of
the Western security system, and would be touched again. The
Korean War demonstrated that "wars of liberation" and the
use of proxy troops could dent the West's defenses outside
Europe. The strategy of containment thus worked but imperfectly; the barriers being erected against Soviet-Communist
expansion could and repeatedly would be outflanked. It would
then fall to the United States to try to improvise hastily a new
set of defenses, to rush into the endangered spot with economic help and military supplies, and in two cases to engage
in large-scale military operations while the USSR stood, so to
speak, on the sidelines, though hardly a disinterested observer.

Apart from disregarding the opportunities which Russia's
postwar vulnerabilities presented for more assertive US policies, the prescription for containment had another flaw: it
preached patience, a virtue seldom exhibited by the Americans when confronted with irksome problems, whether at home
or abroad. The policy, they were told, would accomplish two
purposes: it would keep this country out of war and *eventually* it would make the Russians mend their ways. Both assumptions were rudely shaken by the Korean War. With it
came a belated recognition of the sins of omission incurred by
American diplomacy in the forties. The new Republican administration promised to renounce the sterile containment formula, and Secretary Dulles talked volubly (though once in office with decreasing confidence) about the new guideposts:
"liberation" and "rollback" of the Soviet sphere. Yet such
goals, eminently sensible, say, in 1946, were by 1953 no longer
within the power of American *diplomacy*. They could not be
pursued without incurring a serious risk of a global war. The
USSR was much stronger, not only in tangible industrial and

military terms, but also in what might be called the psychological appurtenances of power. One of the latter was the Soviet-Chinese alliance. Internally, and unbeknownst to the West, this alliance was already suffering from considerable strains. But to the outside world it still stood as a monolithic colossus, the two countries with their combined population of about one billion working hand in hand.

Another Soviet achievement whose psychological impact far surpassed at first its military significance was the acquisition of nuclear weapons. Today, when we have grown accustomed to living in awareness of the enormous destructive potential at the disposal of both superpowers, it is difficult to appreciate the shock produced by news of the first Soviet atomic explosion in 1949, years before anyone in the West thought it would happen, and the even greater agitation in the wake of the Russians' test of their first hydrogen device in 1953. Even then military experts in both countries were well aware that the US still possessed a crushing superiority both in nuclear weapons and in the means of delivery. But psychologically, a small stock of nuclear weapons in the hands of a regime such as the Soviet one tends to offset a much greater number of the dreaded devices held by a democracy. A Gallup poll in September 1945 found that more than 77 percent of the American public approved of the dropping of atomic bombs on Japan and only 5 percent found it totally unacceptable. But during the first phase of the Korean War, when the American forces were pinned down in the southern tip of the peninsula and threatened with destruction, only 19 percent of those polled would have supported the use of nuclear weapons against the North Koreans.[17] Obviously, once the USSR possessed its own (even if rudimentary) nuclear weapons, the American public was more reluctant to see its own government use nuclear force.

By 1953 Soviet prestige was firmly bound up with the preservation of Communist regimes in the Kremlin's satellites. More than that, whoever ruled the Soviet Union now was much more vulnerable domestically on account of his or their stew-

17. John E. Mueller, *War, Presidents and Public Opinion* (New York, 1973), pp. 105, 172.

ardship of foreign policy than before. Stalin's absolute power was in no way diminished or threatened because of Tito's successful defiance of Moscow. But what his colleagues saw as Khrushchev's mismanagement of intra-Communist affairs was undoubtedly one reason behind the plot to unseat him in 1957 and clearly contributed to his overthrow in 1964.

"Liberation," when confronted with the hard facts of American and international politics, was soon revealed as a bracing slogan, rather than a practical policy. Diplomatic pressure backed by tough rhetoric was by now incapable of constraining the USSR to relax its grip on Eastern Europe or to agree to a united and democratic, even if neutralized, Germany. It would be unfair to see the tough talk which emanated from Washington in the beginning of the Eisenhower Administration as having been entirely ineffective. The usual Soviet apprehensions at the change of guard in Washington coincided with the considerable disarray in Moscow following Stalin's death. Fearful of each other, and, at first at least, of their own people, his successors could not emulate the late despot's tactics in bluffing the West and concealing the regime's weaknesses and apprehensions. And indeed, for all of its rapidly increasing military potential and industrial growth surpassing the most optimistic postwar predictions, the Soviet Union of the fifties and sixties would never quite recover that facade of defiant self-confidence and that ability to awe the outside world which it displayed in the immediate postwar period under Stalin. Militarily the Soviet state was growing stronger both in absolute terms and vis-à-vis the US, and yet its political vulnerabilities, and especially those of the world Communist system, were now more discernible to the foreign observer.

Just as they dismantled some of the most oppressive aspects of Stalinism at home, so in their foreign policy the new leaders tried to project a new Soviet image. Instead of implicit threats against the West, they would now stress the need and possibility of peaceful coexistence. After the infrequent and frigid diplomatic contacts of the past, there had opened, beginning with 1954, an era of conferences and summit meetings, a veritable Soviet campaign of cordiality—interspersed to be sure by occasional reminders that, if its peaceful initiatives were spurned, the USSR could protect its interests in other ways.

And there was, to be sure, a modest "rollback." The axiom that the Soviets never relinquished an area they had once occupied was once more confounded by the Austrian Peace Treaty of 1955, which relieved the little country of Soviet military presence and spared it the fate of Germany and Korea. At about the same time the USSR vacated the military-naval bases it had secured in Finland at the end of the war.

Were all such manoeuvres and concessions caused by the tough rhetoric that marked the beginning of the Eisenhower administration? To a degree, yes. But the main cause was that Khrushchev's list of foreign policy priorities was different from Stalin's. The latter had to worry lest the West challenge his imposition of Communism on Eastern Europe, and he tried to prevent the formation of the West German state (which the Kremlin correctly assumed would sooner or later be allowed to rearm). But America's first feeble reaction to the brutal Soviet suppression of the Hungarian revolution must have provided a decisive proof of the hollowness of all the talk about liberation. On the other hand, the German Federal Republic was now a fact of international life, and so after 1955 would be its rearmament and membership in NATO.

The Soviets, though very grudgingly, acquiesced in West Germany's acquiring conventional arms. A German army in the late fifties, when Moscow disposed of a sizeable arsenal of nuclear weapons, was no longer the threat it would have been a few years before. Khrushchev and his colleagues' main anxiety on the European front was now concentrated on two other issues. One was to prevent Bonn from acquiring a nuclear deterrent of its own. The other development they desperately sought to preclude was a close political and military union of Western Europe. United Europe would change completely the configuration of forces on the international scene. It would become the third superpower. And with West Germany assuming, because of its population and economic strength, the leading role in the confederation, it could pose a new threat to the Soviet empire in the east. Hence the twofold imperative for Soviet policy on the Continent: the European Economic Community must not be allowed to evolve into an effective union, and Bonn must be denied atomic weapons and constrained to give up its claim to German unity.

The pursuit of these objectives led Khrushchev to be intermittently a wooer and a bully. At times he exuded cordiality: Russia now (unlike, by implication, in the bad old Stalin days) threatened no one. Hence no need for Western European states to draw closer together, heed Mr. Dulles' fatuous warnings, and waste their money on armaments! Conversely, the Soviet Union was now the world's strongest state, equipped with the most powerful weapons as well as with rockets capable of delivering them accurately on London and Paris. This Jekyll-and-Hyde stance of Soviet diplomacy was most pronounced between 1958 and 1962. Every few months the irrepressible First Secretary would issue a new ultimatum: there had to be a German peace treaty sanctioning the territorial status quo and containing cast-iron guarantees against nuclear weapons for Bonn. If not, the USSR would sign a separate peace treaty with East Germany; West Berlin would be subjected to a blockade with this time no loopholes like an airlift being tolerated. Then, as the Western chanceries were preparing for the worst, the Kremlin would seemingly relent: there really was not that much hurry about the whole business, it could be postponed for a while to allow for another summit meeting or while Khrushchev visited the US, etc. The game failed to achieve its main goals, but it kept the Western alliance off balance and led to serious fissures within it. Britain was most nervous, willing to meet the Soviets halfway; Adenauer's Germany was most intransigent; the US hesitant and uncertain about what Moscow was really after. Was it West Berlin itself, diplomatic recognition of East Germany, or what?

The 1958–62 crisis offers a good illustration of that secretiveness of Soviet decision-making which, while it is usually an element of strength in Soviet diplomacy, proves occasionally to be counterproductive. To state plainly one's objectives, Moscow believes, is to reveal one's fears and thus to give your adversary an edge in any negotiations. What Khrushchev wanted most, an absolute ban on the Federal Republic's independent nuclear deterrent, was something the West would have most likely conceded without asking for any quid pro quo. (Few foreigners could have understood why the USSR, apparently unawed by America's mighty nuclear arsenal, should have been so concerned at the possibility of Bonn acquiring a few bombs of its own.) But the Kremlin did not want to take chances and chose to pursue its objective through a more circuitous and devious route. Hence first the Rapacki plan: the *Polish* foreign minister suddenly developed the inspiration that Central Europe should be a "denuclearized zone," i.e., that atomic weapons should neither be produced nor stockpiled within the several states of the area including, needless to say, the Federal Republic. It would be surprising if as much as a single sentence of the "Rapacki" proposal was written in Warsaw, rather than in Moscow. It was only when the West refused to consider the plan that Khrushchev erupted with his threats about Berlin.

He was equally devious and even more misunderstood in the West when it came to his Chinese policy. The Soviets sought desperately to conceal from the non-Communist world the real state of their relations with Mao's regime. Yet, starting in 1958, they began to look for potential allies, or at least some form of outside help, to enable them to cope with Mao and Co. The key issue was again the bomb. Beijing was determined to develop its own nuclear arms; the Soviets became more and more unhappy over the prospect. By 1959 signs of the growing tension between the two superpowers were already discernible to outsiders, even though the US State Department still kept referring to "the Sino-Soviet bloc." During his visit to the US that year, Khrushchev tried on one occasion to raise the subject of China with Eisenhower, but was rebuffed by the President, who evidently assumed that all the Soviet leader had in mind was to plead for American recognition of the Communist regime.[18]

The deception could not be sustained forever, and in 1960 the bubble of "the unshakeable unity of the Soviet and Chinese

18. Ulam, *Expansion and Coexistence*, p. 626.

peoples" burst amidst mutual accusations now aired in public. But American perceptions of the dispute were at first obscured by the conflicting parties' ideological claims and recriminations. It was but slowly and imperfectly realized in the West that the real roots of the conflict lay in a clash of two militant nationalisms and mutual fears of the two ruling groups.

An incorrect analysis of the momentous split prevented Washington from deriving any concrete benefits from the quarrel between the two Communist giants. On the contrary, and by a strange historical irony, the conflict became one of the main causes of the enhanced difficulties and dangers the United States was to experience during the turbulent sixties.

Unlike Washington, Beijing saw clearly the anti-Chinese impulses behind the Soviet leaders' moves. It thus had every incentive to block any Soviet initiative toward a rapprochement with the United States, as well as to prod Moscow into a more militant stand against the West. For their own part, the Soviet leaders, much as they saw through and resented the Chinese game, still could not remain completely unaffected by it. For one, they were not ready as yet, they felt, for a definitive and permanent break with Mao. For another, Beijing's militant anti-Western rhetoric, unless matched by that of the Kremlin, threatened to undermine Soviet prestige and influence in the Third World. Thus Khrushchev's predicament: he was seeking an accommodation with the US so as to get a handle on the Chinese situation, at the same time that he felt constrained, largely out of fear of being outbid by Mao, to support revolutionary movements and "wars of liberation" which again and again would place him on a collision course with the US. Already in 1958 the Russians felt that their Chinese comrades would have liked to push them into a shooting confrontation with the US.[19] And Beijing's promptings were un-

19. A Soviet source was later quite explicit on the subject: "In August-September . . . Mao and his partisans without consulting the USSR . . . entered upon a provocation, the shelling of islands in the Taiwan straits. . . . They played with the idea of a 'local war' with the US, and the USSR was to be drawn into it at a certain phase." V. I. Glunin and others, *The Recent History of China* (Moscow, 1972), p. 319.

doubtedly instrumental in making the North Vietnamese regime intensify its guerrilla activities in the South in 1960 and
in setting the stage for an all-out campaign to topple America's protégé, President Diem.

In his first official meeting with Soviet Ambassador Dobrynin, Henry Kissinger was astounded to hear the latter declare
that "Great opportunities had been lost in Soviet-American affairs, *especially* between 1959–1963." [20] The then presidential
security assistant told his interlocutor that he could not be
speaking seriously in describing 1959–63 as a period of opportunity: "That was, after all, the time of two Berlin ultima-
tums, Khrushchev's brutal behavior toward Kennedy, the Cuban missile crisis." [21] Dobrynin, possibly feeling he had said
too much, changed the subject. But the incident is instructive
as to the difficulties inherent in communication between the
two superpowers. That barrier, as much as any other reason,
has plagued their relations. One of the most astute American
students and practitioners of foreign policy still failed to recognize that the Soviet ambassador was not merely engaging in
diplomatic small talk: 1959–63 had in fact been a period of
lost opportunities in Soviet-American relations. What Khrushchev sought through his oversubtle policies was also at the
time devoutly wished for in Washington: to curb China and
especially to prevent or delay its emergence as a nuclear power.
To be fair, there is something to be said for Kissinger's retort:
Khrushchev's actual behavior had made it very difficult to see
him as a wooer rather than a bully. But his American counterpart ought to have been more sensitive to the external and
internal pressures under which he was operating.

And indeed the strategic-diplomatic manoeuvre through
which the Soviet leader finally sought to achieve the diver-
gent aims of Soviet foreign policy might well be likened to a
subtle form of blackmail. It was a sudden and brutal act of
force to extort concessions from the US by placing Soviet nu-
clear-tipped missiles in Cuba. To be sure, as Khrushchev probably envisaged, the subsequent scenario of this act of force

20. Henry Kissinger, *White House Years* (New York, 1979), p. 114.
21. *Ibid.*

would include explanations and endearments that were to set the stage for a peaceful and enduring relationship between Russia and America.

There seems little doubt that the missiles were intended to serve as bargaining chips in a complex and many-sided Soviet game. Presumably Khrushchev would have offered to withdraw them from the Caribbean in return for American concessions on (a) Berlin—i.e., a German peace treaty banning nuclear weapons for the Federal Republic and recognition of the German Democratic Republic—and (b) China—again US official recognition of Beijing and the withdrawal of American military protection from Taiwan. China in return, the Kremlin hoped, would refrain from developing its own nuclear deterrent, as well as return to a more cooperative and friendly relationship with the USSR.[22] It would have been Khrushchev who announced the presence of the weapons in Cuba as well as the price for their removal.

The premature discovery of the Soviet installations in Cuba ruined the whole scheme. Instead of pulling off a master coup which could have solved the most grueling dilemmas of Soviet foreign policy, the USSR found itself in a dangerous confrontation with the US. A random accident or a reckless move on either side might well have triggered a nuclear exchange. There followed a humiliating Soviet retreat. The years 1962–64 marked the nadir of Russia's postwar diplomacy: it seemingly ran out of tricks. Threats over Berlin were discontinued. The conflict with China grew more intense, with Moscow now clearly on the defensive, at least on the rhetorical level.

Soviet discomfiture did not lead, however, to any long term gains for the US. As mentioned before, both powers had an interest in containing Communist China: the USSR for reasons which do not have to be spelled out, the US because of the widespread conviction in this country that Beijing preached, and if once industrialized and equipped with modern weapons would practice, a much more aggressive and anti-Western

22. A more detailed reconstruction of the Cuban missile crisis is presented in my *Expansion and Coexistence*, pp. 667–76, and in *The Rivals*, pp. 299–340.

brand of Communism than Moscow's. But after 1962 it was no longer within Khrushchev's power to alter the basic lines of Soviet foreign policy. The only notable result of what might be called the first, and brief, Soviet-American detente (1962–65) was the nuclear test ban agreement, signed by the US, the USSR, and Britain in July of 1963. One year later an event occurred which both Washington and Moscow would have paid dearly to prevent: China exploded its first atomic bomb. Rather symbolically, about the same time Khrushchev's political career came to an end.

By 1964 both superpowers had come to view their mutual relations, as well as the general world situation, with a degree of resignation. On the US side there was now a realization that there was no magic formula to conjure away the Soviet challenge. "Containment," "liberation," "negotiating from a position of strength," and similar incantations had proved of little use in coping with such troublesome developments as Soviet forays into the Third World or in countering Moscow's exploitation of disarray and conflicts consequent upon the dismantling of the Western empires. Direct negotiations and summitry were hardly more effective; those "spirits" of Geneva and Camp David proved short-lived and bereft of any practical results.

On the other side, Khrushchev's ouster put an end to foreign policy through pyrotechnics—spectacular bluffs, threats, and cajolery, or trying to resolve the outstanding differences between the two superpowers by a single master stroke. Brezhnev and Kosygin's list of foreign priorities was not different from that of their predecessor: to seek both a reconciliation or at least papering over of the differences with China and a peaceful settlement of the German problem. But now these aims would be pursued more prudently and patiently than heretofore. The language of Soviet diplomacy would become less exuberant. No nuclear missile-rattling, as under Khrushchev when "in order to appear willing to take risks they knew to be imprudent, the Soviet leaders . . . deliberately conveyed an exaggerated impression of the size and scope of their ICBM program." [23] At the same time the Kremlin embarked upon a more systematic and ambitious arms build-up, its main emphases being to catch up with and to overcome the US in strategic weapons, and to develop a powerful ocean-going navy. The stepped-up tempo of the build-up was to prepare for all sorts of contingencies, such as a drastic turnabout in US–Communist China relations. As yet such a possibility would have been dismissed as sheer fantasy by Washington. But as early as February 1964, Politburo member Michael Suslov, speaking before the Central Committee, attacked the Chinese leaders and warned that "they would not refuse to improve relations with the United States but as yet do not see favorable circumstances for such an endeavor." [24]

For all of its impressive industrial growth of the previous two decades and consequent advances in missile technology and nuclear weapons, the Soviet Union's overall international position as of the end of 1964 appeared unenviable. The goal which had hitherto eluded American policy-makers—a situation in which Russia's main avenues of political and ideological expansion would effectively be barred—was close to being realized. This new configuration of world politics was not due to some master stroke of American diplomacy, but rather to two developments, one for which the US could claim no credit, and another to which indeed it had made a major contribution, though economic rather than political in its inception.

First, the Sino-Soviet split put in question the whole rationale of the Kremlin's efforts to expand its power and influence by supporting revolutionary and "national liberation" movements. Now the two Communist powers were in a way trying to contain each other, and China's anti-Russian campaign was especially damaging to the Soviet image and influence in the Third World.

The other development—from the Soviet point of view so-

23. Arnold L. Horelick and Myron Rush, *Strategic Powers and Soviet Foreign Policy* (Chicago, 1966), p. 119.

24. *Plenum of the Central Committee of the CPSU, February 10–25, 1964* (Moscow, 1964), p. 495.

bering, for the US encouraging—was taking place in Western Europe. The great success of the Common Market and the burying of such ancient enmities as between France and Germany created a momentum toward political unity of the area. For the moment the formidable figure of de Gaulle appeared to be the main obstacle to this. But the logic of the situation urged that even if the aged leader should continue in power for several more years, the impetus toward a political union, and one including Great Britain, could not be arrested. Such a union in turn would reverse the paradoxical phenomenon of the postwar years when, though fully recovered, then flourishing economically, the Western European nations had steadily declined in political importance. United, Western Europe would in every respect become the third superpower, and as such would restore the politico-military balance on the Continent and stop the erosion of Western influence elsewhere.

The combined effects of the Sino-Soviet dispute and Western European economic strength were already being felt, even within the Communist bloc. The reverberations of the intra-Communist conflict enabled Albania to slip entirely from the Soviet grasp and helped Romania to achieve a degree of freedom in its foreign policy. And most of the people's democracies were now beginning to look to the West for profitable trade and credits. Were the trend to continue, the Soviet Union was bound to find itself under considerable pressure to loosen its controls over the satellites and to allow them to liberalize their internal policies.

All in all the difficulties and constraints encountered by the Soviet Union in 1964 appeared to be more serious than any it had experienced since Stalin's death. A Western observer would have had good reasons to view the picture optimistically; it promised "to force upon the Kremlin a far greater degree of moderation and circumspection than it has had to observe in recent years" and to bring about that "gradual mellowing of Soviet power" which Kennan had defined as the aim of containment.[25] As on previous occasions, such a fortunate turn of events would have also required prudent and pur-

25. Kennan, "The Sources of Soviet Conduct," p. 124.

poseful actions on the part of the US. Careful reading of the facts and figures of the international scene would have urged Washington to give the highest priority in its foreign policy to the promotion of Western European unity, as well as to eschew any action that might obscure the significance of and divert the world's attention from the Sino-Soviet conflict.

On both counts, not to mention other considerations, the US intervention in Vietnam proved to be a momentous blunder. It weakened this country's standing and ties with its European allies, and thus undercut its ability to prod them into a more effective political and military alliance. They would now seek to deal with the USSR separately, the ephemeral visions of de Gaulle's "Europe from the Channel to the Urals" and Brandt's *Ostpolitik* handing Soviet diplomacy new opportunities and advantages. And if in 1964 the Kremlin was on the point of seeking a far-reaching accommodation with the United States, then Vietnam greatly diminished both its ability and incentive to do so. Moscow's political and material support for Hanoi enabled it at no risk and at relatively little cost to refute Chinese charges of Soviet-American collusion, and in general to regain within the international Communist movement much of the ground it had lost to Beijing.

With America's attention and its diplomatic and military resources increasingly absorbed by the indecisive conflict in Southeast Asia, the USSR could and did push its efforts at political penetration of the Middle East, Africa, etc., with new confidence. The drama of Vietnam served to obscure and minimize the negative effect of the Soviets' miscalculations—such as spurring the Arabs to their disastrous 1967 confrontation with Israel—just as the US overcommitment in Southeast Asia probably made it easier for the Politburo to decide to invade Czechoslovakia in 1968. By the end of the 1960s it was still difficult to predict when and how the conflict in Indochina would end, but there already could be no question as to who had been its main beneficiary. It was the USSR.

Almost a quarter of a century before, expounding the doctrine of containment, Kennan had asserted that "The United States has it in its power to increase enormously the strains

under which the Soviet Union must operate." The US involvement in Vietnam had the opposite effect; it helped free the
USSR from the many strains that impinged on its foreign policies in the mid-sixties and that, if properly gauged by Washington, might well have led to that "gradual mellowing of Soviet power" hoped for by the exponents of containment. As it
was, America's predicament in Southeast Asia, on the one
hand, and the general disarray of Communist China, on the
other, strengthened Moscow's hand against both its rivals. As
against their domestic turmoil, the USSR appeared a haven of
tranquility. Washington's foreign policy was crippled by
preoccupation with the intractable dilemma of Vietnam, Beijing's virtually paralyzed by the chaos into which the Cultural
Revolution had thrown not only the government machinery
but society as a whole. By contrast the Kremlin's foreign image was, at least for the time being, one of prudence and
steadfastness. It had been helping Hanoi, and could claim that
it had enabled a small Communist nation to withstand the
pressure of the American colossus while at the same time
avoiding an actual Soviet-US confrontation. In the 1969 border clashes with China, the Soviets displayed both firmness
and restraint. Vietnam and Mao's folly helped blot out the
memories of the Soviets' ineffectiveness in protecting their
Arab friends in the Six-Day War and their brutality in suppressing Czechoslovakia in 1968.

Nevertheless this favorable configuration of international
events could not continue indefinitely. The US sooner or later
would extricate itself from Vietnam. Mao's regime would
eventually have to put an end to the contrived anarchy into
which it had plunged China. It then might—there were already signs that it would—seek a *rapprochement* with the US.
Thus at the end of the 1960s the Soviet leadership had every
incentive to cash in on its gains of the last few years and to
start bargaining with the West. It was now in a good position
to achieve what had eluded both Stalin and Khrushchev: a
settlement of the German problem and official recognition of
the postwar Central and Eastern European frontiers. Perhaps
it could no longer hope to bar the road to an American-Chinese
rapprochement, but it still could limit its scope and potential

damage to Soviet interests. The USSR had now fewer reasons to fear any untoward domestic consequences of a freer economic and cultural intercourse with the West. It was no longer the devastated and impoverished society of the postwar era. The Kremlin could now negotiate from a position of strength. In 1945 Russia's production of steel had been less than one-sixth that of the US. By 1970 the respective outputs of this commodity most indicative of a society's industrial strength were virtually equal. When it came to strategic weapons, America's once absolute and then for a long time decisive superiority was also a thing of the past. In 1969 the number of Soviet ICBMs stood at 1050, as against 1054 for the United States.[26]

A different but equally cogent set of reasons urged the new administration in Washington to seek a new opening in Soviet-American relations. President Nixon came into office committed to finding an honorable conclusion to the Vietnam war. The key to such a settlement, it was not unreasonably believed, lay with Moscow. Apart from Vietnam there were other factors which strained dangerously that national consensus on foreign policy that had endured for two decades, spawning American commitments, bases, and aid missions all over the globe, making the US taxpayer acquiesce in literally hundreds of billions being spent not only on our own armaments but also on military and economic assistance abroad, and allowing young Americans to be drafted in peacetime. Some twenty-odd years before, one could rationalize all such burdens as a price the American people had to pay for "accepting the responsibilities of moral and political leadership that history plainly intended them to bear."[27] Alas, by the late 1960s a great number of Americans had grown skeptical as to whether history had indeed predestined this country to assume such vast responsibilities and bear such heavy burdens. One practical reflection of this skepticism was the increasing reluctance of Congress to authorize expensive new defense systems and weapons that the Pentagon asserted were needed to keep

26. Institute of Strategic Studies, *The Military Balance, 1968–69* (London, 1970), p. 54.

27. Kennan, "The Sources of Soviet Conduct," p. 124.

the US ahead of the Soviet Union in the strategic arms race. That race itself was now seen by many, including some in the government, as both futile and monstrously expensive for an economy increasingly plagued by inflation. Both superpowers had enough weapons to destroy each other several times over. Simple logic pointed out the folly of piling up more and more weapons. Whether what was obvious to the average American was equally so to a Soviet marshal or Politburo member was at the time not widely considered.

A strategic arms limitation agreement thus seemed the most logical and decisive step in the endeavor to change the current and unfortunate pattern of Soviet-American relations.

At first glance it appeared a hopeful sign that the cornerstone of the proposed *rapprochement*, both parties agreed, should be a treaty which could be phrased in concrete technical and quantitative terms; an issue to which ideological incantations whether of the "free world" or "scientific socialism" variety bore no relevance, an agreement aimed at minimizing the possibility of a calamity so colossal that compared to it all other problems of world politics faded into insignificance.

Someone who had followed the world scene since the end of the war should at the time have recognized the risks of placing an *arms* agreement on the very top of the agenda of Soviet-American negotiations. Such an agreement would have to recognize virtual nuclear parity between the two powers: in a sense this would be but an acknowledgment of the facts and figures of the international situation as of 1970. But symbolically and psychologically this acknowledgment would be read by many throughout the world as registering a decline of America's power and of its ability to preserve what was left of the international order. Even at the height of its economic and military power the US could cope but imperfectly with Soviet expansionism and growing anarchy in international relations. The new era opening in Soviet-American relations still held a great promise for both countries and the world, but only if the US could learn new political and diplomatic skills and the Kremlin unlearn some of its old habits.

2

The Making of Detente

As their country entered the seventies, the rulers of the Soviet Union were in a position to appreciate the uncomfortable discovery made by so many Americans in the forties and fifties: enhanced power does not automatically, especially in this nuclear age, give a state greater security. From almost every point of view, economically and militarily, in absolute and in relative terms, the USSR under Brezhnev was much more powerful than it had been under Stalin. And yet along with this greatly increased strength came new international developments and foreign commitments that made the Soviet state more vulnerable to external danger and the turbulence of world politics than it had been, say, in 1952.

The Communist regime and the Soviet empire emerged successfully from the travails of de-Stalinization. The Kremlin's rule was firmly established over Eastern Europe. But this domination carried with it the burden of constant vigilance and readiness to intervene in the internal affairs of six nations with a combined population of over 110 million—witness Czechoslovakia in 1968. (And the assumption that the suppression of the "Prague spring" would immunize Eastern Europe from serious trouble for at least several years was soon shaken by the Polish events of December 1970.)

The USSR had also established its presence in several areas that until quite recently had been the exclusive preserve of the West, notably the Middle East. As a result the Soviet Union's prestige was now closely involved in the affairs of its Arab clients and their seemingly endless confrontation with Israel. It has taken some very fancy diplomatic footwork on the part of the Kremlin to erase the impression produced in the wake of the Arabs' defeat in the 1967 war that Moscow was less than whole-hearted in standing behind its Middle Eastern allies and protégés in their hour of need. The Soviets did not desist from trying to maintain their influence in the area, for nowhere else outside of Europe were the West's vital interests so seriously engaged and the US position so vulnerable. By 1970 America's resolve on Vietnam was growing ever weaker, and the Soviet Union was getting credit for exactly the right character and amount of help to Hanoi and its friends. Yet only a few years before, following the first American bombing of the North, the Soviet leaders had been criticized within the world Communist movement (and not only by the Chinese) for not taking stronger and more direct measures to protect North Vietnam and to advance the Vietcong cause in the South.

Another reason why Brezhnev's Russia, while immeasurably stronger than Stalin's, could not feel equally secure and confident about the future was what by now has become an irreversible split within world Communism. Both Moscow and Beijing still continued to advance ideological explanations for their conflict, but it was becoming increasingly clear to the outside world, as it undoubtedly had always been to the respective leaders, that differing interpretations of Marxism-Leninism had but little to do with the real roots of the dispute, which were to be found in the natural incompatibility of two intensely nationalist and absolutist systems. The excesses and ravages of the Cultural Revolution had, as we have already noted (and as Mao's successors would themselves acknowledge in a few years), dealt a serious blow to Communist China's development, as well as Beijing's prestige within the world

Communist movement. But for all its self-inflicted wounds, Mao's regime still remained the master of the world's most populous nation, and still resolutely anti-Soviet. The interplay of fanaticism and hard-boiled pragmatism so characteristic of Chinese Communism was now to receive a new twist: hardly did the regime decide to terminate the folly of Cultural Revolution than it began to signal its intention to seek a *rapprochement* with the US. Though not unexpected by the Kremlin, this was a dizzying turnabout insofar as Western and especially American public opinion was concerned: as recently as the late sixties, China was considered in the US as by far the more hostile of the two Communist powers, and its leaders believed to be Communist fanatics as against their more pragmatic (or, depending on one's preference, cynical) Soviet counterparts.

For the Russians the impending shift in Beijing's policies threatened to add another and most disturbing dimension to their rivalry with the US. In its official rhetoric the Kremlin always identified America, or rather its "ruling circles," as the main threat to the USSR as well as "socialist and progressive forces" throughout the world. But in fact the Soviet leaders had always recognized that the realities of American politics, as well as a democracy's absorption in its internal problems, acted as a powerful barrier to Washington's developing militantly anti-Soviet policies and minimized chances of the rivalry of the two superpowers leading to an armed confrontation. Barring an unusual set of circumstances, as in the Cuban missile crisis of 1962 (blamed by Khrushchev's successors on his recklessness, which they were determined to eschew), no American leaders, even those as anti-Soviet as the late Secretary Dulles, had been able to propel American policies into a course of action that threatened a nuclear conflict or presented a basic and effective threat to the Soviets' vital interests. But there was one important qualification to this rather reassuring (even if in some ways unflattering) assessment of American policies and their makers. In the past, the Kremlin believed, a third party had frequently played the role of the troublemaker insofar as relations between the two superpowers were concerned, pleading with and occasionally convincing the American colossus that it must stand up to the Russians. Britain's Churchill and Bevin helped to inspire and put into effect the policy of containment, Adenauer's Germany in the fifties fortified the Eisenhower Administration's refusal to recognize formally Soviet hegemony over East Europe. Of late the situation had been drastically reversed: with the growth of Soviet power and relative decline of America's, it was the Europeans who now pressed for improved Soviet-American relations. Though ritualistic references to "the Bonn revanchists" still abounded in the Soviet press, the new Chancellor, Willy Brandt, had in effect scuttled Adenauer's policy of uncompromising opposition to the division of Germany, and was working toward a detente with the Soviet bloc.

America's European allies were thus becoming more accommodating to Soviet wishes, and the US, chastened by Vietnam and its domestic troubles, appeared ready to follow their example. Yet precisely at this point the prospect of a *rapprochement* between the People's Republic and the US raised the unpleasant possibility of a new devil's advocate—Beijing—pleading with Washington not to let its guard down and to persist in its distrust of Moscow's promises and pledges. Mao's regime, the Kremlin realized only too well, was not seeking to open a dialogue with the US just for its own sake, or even in order to remove America's protective shield over Taiwan. Important as the latter objective was for the Chinese Communists, their main concern was still to try to prevent a *rapprochement* between the two superpowers. And to prevent or wreck Soviet-American detente, Beijing would not hesitate to enter into a virtual alliance with the main capitalist power.

In 1970 the fears of both sides would have struck an outside observer as excessive, if not indeed bordering on paranoia. How could one envisage Moscow and Washington establishing a joint consortium to rule the world, or "co-hegemonism," as the Chinese Communists dubbed this fantastic prospect? Equally fanciful must have appeared the Kremlin's notion, or rather nightmare, of a Sino-American alliance. As of the moment the Nixon Administration's goals concerning China did not go beyond a modest degree of normalization of relations between the two countries. Rather than being part of some nefarious anti-Soviet plot, such normalization, it was hoped, would in turn help in reaching a satisfactory accommodation with the USSR. The very nature of America's political system and the popular mood placed severe constraints on Washington's ability to play off one Communist power against another. The President's national security assistant, Henry Kissinger, was indeed known as a student and admirer of nineteenth-century European diplomacy. But whatever the personal inclinations and talents of US policy-makers, not even a Talleyrand or a Metternich with their intricate schemes could have gotten very far had they been closely watched by an irascible Senate and the inquisitive media, and placed in a society where the very term "power politics" was viewed reprovingly and believed by many to connote some kind of black magic. In brief, America's current mood made the chances of its joining the Kremlin in an anti-Chinese intrigue as minimal as those of a Sino-American alliance directed against the Soviet Union.

Taking a longer view, however, there was a substratum of realism in Moscow's and Beijing's suspicions of each other. In his intermittently clumsy and oversubtle ways Khrushchev had probed whether America could help him to contain China. China's virulent anti-Americanism before the 1970s had been largely designed to bar any reconciliation between the USSR and the US, if not indeed to provoke a confrontation between the two. The Brezhnev-Kosygin team tried to mollify the Chinese, to make them at least join in papering over the dispute, but with no success. The Cultural Revolution lessened the seriousness of the Chinese challenge but brought in its wake fresh dangers. The Kremlin had never been quite comfortable with inflammatory Maoist rhetoric, e.g., gleeful predictions, both before and after the split, of an inevitable war between the US and the Soviet Union, and rather ghoulish boasts that the Chinese themselves were not afraid of a nuclear conflict since several hundred million Chinese would survive, while presumably the capitalists and the revisionists finished each other off. But, having used similar though much less drastic scare tactics vis-à-vis the West, Moscow, though with some misgivings, had tended to discount the possibility that Mao and Co. really meant those horrendous predictions. Still the recent Chinese extravaganza—when the aged and possibly unbalanced despot had let loose the Red Guards, bands of young fanatics, to wreak havoc with the country's whole political structure—must have led to some rather anxious reflections in the Kremlin. Even Zhou Enlai, whom the Soviets (quite rightly) considered the main force for reason and moderation amidst all the madness, was at one point besieged and confronted in his office by the Red Guards. From Brezhnev and his colleagues' perspective, the whole picture would have been highly amusing except for the following uncomfortable facts: (1) China and the USSR have the world's longest common border; (2) though technologically still quite backward, the People's Republic disposed of virtually limitless military manpower, and (the biggest fly in the ointment) already possessed a small stock of nuclear weapons.

It was in 1966 that the Soviets began to concentrate sizeable combat units equipped with tactical nuclear weapons on the Chinese border. It is also quite likely that concern with the behavior and potential moves of their erstwhile ally and protégé was one of the main reasons why the USSR accelerated its strategic as well as conventional arms build-up in the mid-sixties. The number of Russian ICBM launchers rose from 250 in 1966 to 570 in 1967 and 900 in 1968.[1] A mainly political and partly ideological challenge was fast becoming a military one as well. For the moment the potential military threat Beijing posed to Soviet Asia was not serious, that is if one could discount some insane and suicidal nuclear provocation. China, its economic backwardness aggravated by the reverberations of the Cultural Revolution, had a long way to go before it could become a major industrial power capable of supporting modern military technology. It had acquired a hydrogen bomb, but as yet had no facilities for the production of modern aircraft. But turning a new leaf in its relations with the US would give

1. Figures in Henry Kissinger, *White House Years* (New York, 1979), p. 537.

it access to Western technology. Though it had already opened trade with the Western European states, China's poverty had hitherto stood in the way of speedier and more efficacious industrialization. So did Mao's principle of self-reliance, which precluded seeking extensive credits from the capitalist countries. But it was unlikely that such ideological scruples would continue to prevail for long, especially once Mao was finally gathered to his ancestors, and Washington and Beijing found themselves in a new and friendly relationship.

As of the beginning of 1970 there had been as yet few concrete signs of such a development, and it was the US that believed it had to act the role of the wooer. American restrictions on trade with and travel to China had been relaxed in July 1969. In December of the same year those rather bizarre meetings between the two countries' representatives in Warsaw, suspended because of the war in Vietnam, were renewed. Held intermittently since 1955 (and presumably scrupulously monitored by Soviet intelligence agents), they had hitherto been entirely barren of any concrete results and seemed unlikely to produce anything sensational in the future. Still, all these were straws in the wind, and Moscow felt constrained to show its hand more than it usually cares to do in such situations. In January 1970 Soviet ambassador Dobrynin asked the State Department to keep him posted about the content of American-Chinese communications,[2] a gentle hint to the Americans rather than a case of genuine curiosity; if not Dobrynin, then his Moscow bosses must, for the reason spelled out above, have had a fairly accurate record of what was being said between the two parties. The ambassador's inquiry was thus a hint that Moscow was closely watching what the Americans thought was their subtle wooing of the Chinese but what Moscow considered a coy, but still shamelessly unprincipled, flirtation by Beijing.

There was a fourth capital vitally interested in what was going on between the two Communist powers and America. It was not difficult for Hanoi to perceive one of the main reasons behind the US diplomatic gambits, the desire to disengage it-

2. Kissinger, op. cit., p. 524.

self without losing face from its ill-fated Indochina enterprise. Moscow and/or Beijing, Washington hoped, would pressure North Vietnam to reach a political settlement of the war in the South, one which would allow the Nixon Administration to claim that it had obtained peace with honor. But what might have been a reasonable expectation in 1966 or 1967 was by 1970 a forlorn hope. What had happened in the meantime was the decisive turn of American public opinion against the continuing American military involvement, which had compelled the administration to negotiate while gradually reducing American troops in the South. Soviet diplomacy thus had good reasons to be helpful and understanding of America's predicament, but only up to a point: the Russians used their good offices to arrange secret talks between Hanoi's representative and Kissinger, but insisted that they could not dictate to their North Vietnamese friends what they should or should not do. At the same time, they kept hinting that an overall US-USSR agreement would undoubtedly have a salutary effect on the Indochinese situation as well.

The most obvious aspect of detente, in fact for many throughout the world an issue virtually synonymous with it, was a strategic arms agreement between the two superpowers. For a generation the world had lived with the specter of a nuclear war and the realization that the new weapons and techniques of destruction dwarfed in their power and quantity those used on Japan in 1945. Therefore it was, and is, understandable that any agreement on the dreaded weapons between the two powers that now possessed them in such huge quantities had long been considered a prerequisite not only of a US-Soviet detente, but of a new and better era of international relations. Rationally, one could challenge this stress on SALT as being *by itself* the key to peace and international security. Weapons themselves don't cause wars; conflicting policies of great powers do. Any possible treaty delimiting the number and character of missiles, warheads, etc., to be possessed respectively by the US and the USSR would still leave each with enough to devastate the other. But psychologically, and especially for public opinion in the democratic countries,

SALT had taken on a meaning quite apart from any conceivable content of an actual treaty; its mere existence seemed to
offer a measure of assurance against the unimaginable horror
of an all-out war.

To be sure, in the United States this relief at the prospect of
SALT clashed with a measure of frustration in that it would
acknowledge the end of US nuclear superiority. And among
this country's European allies, satisfaction over the prospective arms control agreement and resultant general detente was
also tempered by a sobering thought. With America's superiority in strategic weapons gone, would it still be able to extend a credible "nuclear umbrella" over the West? Predict-
ably, once the agreement was actually reached in 1972, there
was general relief, followed by almost equally widespread
complaints and fears: in the US that Nixon had "given away"
America's superiority, in Western Europe that this country
would now be loath to risk a nuclear attack on its own territory and thus might acquiesce in its allies being cowed by the
Soviets' powerful land forces.

Such vagaries of public opinion in the West were well appreciated in Moscow. The Soviet Union itself could negotiate
SALT free from the variety of political and psychological pressures that inevitably affected the position of the United States.
The Politburo did not have to contend with proponents of unilateral disarmament, nor with arguments of the "better to be
under capitalism than dead" variety. Some Soviet military men
indeed might have felt that it was a mistake to agree to any
limitation of numbers or character of nuclear weapons when
the USSR was gaining rapidly on the US, at least in terms of
quantity. The American land-based long-range (ICBM) missile
force had been frozen in 1965 at 1054, and at the pace the
Soviets were adding to theirs (from 570 to 900 to 1050 in the
consecutive years 1967 to 1969) they would soon be far ahead,
especially since Congress had recently been unwilling to authorize new and expensive weapons systems. Yet, as the Politburo had good reason to recall, the Americans were capable
of surprises: their usual mood of indecision and acquiescence
in the face of Soviet advances would occasionally be punc-

tured by a sudden and frightening outburst of energy. For the moment Washington was adhering to the doctrine of MAD (mutual assured destruction), which proclaimed that beyond a certain point, i.e., when you are sure to have enough weapons to pulverize your enemy, it is unnecessary and wasteful to produce more, since what you already have is sufficient to deter a strike by the other side. The Soviet Union's compulsive accumulation of more and more ICBMs was thus viewed by the US strategic-doctrine sages with condescending amusement: it was all a lingering reflex of the Russian feeling of insecurity in the past when they were so inferior in strategic weapons, and very soon they too would realize the compelling logic of MAD and stop wasting their money. But the Americans had been known to change their strategic doctrine; their complacency over the Soviets gaining on them might be replaced by alarm and frantic efforts to regain the lead.

Moscow's attitudes concerning the possibility of an all-out war, nuclear weapons, and other related questions have always been more nuanced than those of Washington. First of all it was believed in the Kremlin that a democracy such as the US found it difficult to exploit its nuclear power for political purposes—a conclusion, as we have seen, justified by the experience of the previous twenty-five years. Officially Soviet propaganda on this subject followed a two-fold theme: the capitalists would like to frighten others with their weapons, but they really don't know how. As a Soviet source put it, "During the period 1946–75 the US has threatened to use nuclear weapons on thirty-three occasions (twice against the USSR), but has never actually resorted to them, realizing the obviously catastrophic consequences of a nuclear war."[3] The figure is clearly contrived; no actual instances of such threats are cited; but the implication is clear: much as it would like to, the US does not know how to practice nuclear blackmail. For the bulk of its citizens (the above work, with its small

3. Institute for the USA and Canada of the Soviet Academy of Sciences, *Global Strategy of the USA under the Conditions of the Scientific Technological Revolution* (Moscow, 1979), p. 68.

printing, was obviously intended mainly for the official and professional elites) the Kremlin's message has always been even more reassuring: the great and growing might of the USSR, the influence of the "peace-loving forces throughout the world," keep in check those elements in the American ruling circles who would like to indulge in nuclear blackmail or even resort to war.

It is hard to gauge how far such officially propagated themes corresponded to what was actually thought in the innermost councils of the Kremlin. Khrushchev's successors abandoned his uninhibited rocket-rattling (and it would be instructive to calculate on how many occasions he—quite explicitly and publicly—had threatened to use nuclear weapons). It is clear that the Cuban missile crisis shook the Soviet regime out of its relative complacency concerning America's superiority in strategic arms and was an important factor in its rapid build-up of its own forces in the years following the dramatic confrontation of October 1962. Still, and despite all that their own propaganda had been trumpeting on the subject, the Soviet leaders retained a rather touching trust in the US: barring an unusual provocation, this country was not likely to initiate a nuclear conflict. Strange though it may sound, as of 1970 the Kremlin was more worried by China's still small stock of strategic weapons than by America's vast one.

Some years earlier there had been a discussion of the possibility of West Germany acquiring nuclear weapons of its own. A leading Soviet authority on international relations painted a somber picture of the potential consequences of such a development. "The Federal Republic having received nuclear warheads, could, keeping it secret from the United States, equip one of its vessels with a nuclear missile. . . . Then it could send it close to the US shores to carry out a strike against its territory in a manner suggesting that the attack came from a Soviet ship or submarine. . . . The US government, which would have to decide upon retaliation within minutes, might then order a nuclear strike against the USSR. Be it even . . . a single missile striking one city . . . the USSR would inevitably have to retaliate. The adventurers' goal to provoke a major nuclear war, would thus have been achieved."[4]

West Germany eventually acceded to the nuclear nonproliferation treaty of 1968. But if in the preceding passage, one replaces "the Federal Republic" by "China," one begins to have a fuller picture of the concerns that have dictated Soviet policies on strategic arms. It was not only competition with the US, but also—perhaps we would not exaggerate by saying mainly—the direct and indirect threat posed by the Chinese acquisition of these arms that prompted the Kremlin to (1) step up the pace of its own nuclear arms build-up in the 1960s, and (2) to seek SALT talks with the US. As against their dilatory tactics on the issue in the fifties, and Khrushchev's attempt to tackle the problem in the early sixties through trickery,[5] by the end of the decade the Soviets were more than anxious to negotiate and reach an agreement with the Americans. The nonproliferation treaty signed on July 1, 1968, provided some, though not a very substantial, measure of comfort for the USSR. The vast majority of the signatories were currently technologically incapable of producing a jet engine, let alone a nuclear warhead. What was more reassuring was that the US and Britain joined the USSR in the pledge not to transfer atomic weapons or technology relevant to them to third parties. SALT talks between the two superpowers followed in the wake of the treaty and were expected to reach their climax with President Johnson's visit to Moscow scheduled for September 30. It is interesting to speculate about the possible consequences if SALT I and Soviet-American detente had materialized some four years before they actually did, at a time when few in the United States even envisaged the possibility of a *rapprochement* with Beijing. Would the Russians have been more candid in spelling out their fears about China, and the Americans more willing to listen to them sympathetically? Perhaps the Soviet Union in turn might have been more inclined to help the US extricate itself from the Vietnam imbrog-

    4. H. A. Trofimenko, *The Strategy of Global War* (Moscow, 1968), p. 229. My italics.

    5. See the preceding chapter, p. 31.

lio. Amongst the subjects proposed by Moscow for the agenda of the summit was study of "steps to rule out the accidental appearance of conflict-fraught situations involved in the use of strategic arms."[6] It is not difficult to decipher this cumbrous language: from the Soviet point of view, Communist China was the main instigator of actual and potential "conflict-fraught situations."

It was another intra-Communist crisis that led to the cancellation of the Moscow summit. The Soviet invasion of Czechoslovakia in August made it impossible for President Johnson to become the first American president to visit Russia in peacetime. SALT receded, and when American-Soviet talks resumed, it would be under a new administration in Washington and in a greatly changed international situation.

The SALT process once more became protracted. Technical discussions between the two countries started again in November 1969 and were held intermittently during the next two and a half years in those two "unaligned" capitals, Vienna and Helsinki—a choice unwittingly symbolic of the alternative moods of hope and discouragement that have characterized American-Soviet relations. Austria was the unique example not only of the great powers having terminated their military occupation of a country by mutual agreement (in 1955), but also of both observing scrupulously its neutrality and leaving the fortunate nation to its own devices. Finland's status was on the contrary a reminder of the Soviets' tenacity in holding on to at least some appurtenances of their conquests; it was allowed internal freedom, but in its foreign and defense policies it had to heed its powerful neighbor's wishes. The two teams of negotiators with their scientific and military experts had to wrestle with a bewildering variety of intricate technical problems, the awesome technology of destruction being as it were a third party to their talks and defying human ingenuity to comprehend, let alone control, its ever-changing and more frightful devices. Partly resembling a science-fiction fantasy where machines begin to control men, the process of

6. Gerard Smith, *Doubletalk: The Story of the First Strategic Arms Limitation Talks* (New York, 1980), p. 21.

strategic arms negotiations also exuded a strange kind of scholasticism: experts had to wrestle with defining and distinguishing concepts such as "parity" and "equivalence"; the argument whether it was possible to have a *limited* nuclear war (i.e., one in which nuclear weapons are used only in the theater of operation) sounds somewhat reminiscent of the ancient theological dispute as to the permissibility of belief in purgatory. And no conundrum perplexing a medieval schoolman could have been so difficult as the task laid upon both delegations by one of the directives for the negotiations: they were to devise a scheme "so balanced that neither side could obtain any military advantage, and that equal security should be assured for both sides." [7]

Those intricacies of American-Soviet negotiations were entirely ignored by a third party. In its traditional pre–*rapprochement* with Washington rhetoric, Beijing saw the whole business as further sinister collusion between the two superpowers: "The US-Soviet talks on so-called 'strategic arms limitation' are aimed at further developing their nuclear military alliance. They mainly hope to maintain a nuclear monopoly and carry out nuclear blackmail by nuclear threats against the Chinese people and the people of the world." [8]

Such straightforward analyses might well have led to some discreet sighs in the Kremlin: if only the Americans could see the impeccable logic and benefit to themselves of proving the Chinese Communists' allegations right!

In fact the two allies of Beijing's contrived fantasy were as of 1970 on opposite sides of several major world conflicts. In Vietnam the Soviets' military supplies and other help to the North were crucial in its ability to continue the war; Soviet-made ground-to-air missiles were bringing down American planes. In the Middle East Moscow felt constrained to go beyond supplying military materiel in order to help its protégé. The Soviet position there had been severely shaken in the wake of what the Arab states, and especially Egypt, considered Moscow's insufficient support of their cause in the 1967 war. The

7. Smith, *op. cit.*, p. 21.
8. Quoted in *ibid.*; p. 35.

Soviets thus felt unable to refuse Cairo's request for manpower, as well as equipment assistance, in the intermittent
armed clashes with Israel that flared up along Suez in 1969
and in which Israel's pilots (flying American-built planes) had
been able to raid the Egyptian hinterland almost at will. Some
12,000 Russian troops were now manning Egypt's missile sites,
and Soviet pilots were flying combat missions. The cease-fire
agreement of August 7 ended momentarily the danger of escalation of the confict.

The Middle East presented a good example of the grueling
dilemmas of Soviet expansionist policy. There had been no
good reason for Moscow to engage itself so extensively on the
Arabs' side against Israel (which the USSR was the first to
recognize officially upon its birth) except that by doing so, the
Soviets were gaining a foothold in an area of vital importance
to the West. The risks and costs to the Soviet Union of this
commitment have been formidable; *direct* returns, in terms of
such traditional dividends of imperialism as territory and economic benefits, nonexistent. Yet once there the USSR could
not disentangle itself from the Middle Eastern scene or reduce
substantially its commitment without a serious loss of prestige, and hence reverberations in other parts of the world. Thus
instead of manipulating its clients' policies, it found itself occasionally being manipulated by them, as in the instance just
mentioned. And the Soviets' hopes of safeguarding their influence in Egypt suffered a definite setback with Nasser's death.
His successor as President and national leader, Anwar Sadat,
saw, and not unreasonably, Soviet policy as consisting in
keeping the area in a state of continuous and controlled turmoil, rather than in a genuine commitment to the Arab cause.
In 1971 Sadat purged his military and political apparatus of
those believed to have close ties to Moscow. In 1972 the Soviet "advisers," some 20,000 strong, were politely invited to
leave the country. As long as he continued to preserve a militant stand against Israel, the Russians would have no alternative but to support Sadat. By the end of the decade, Egypt
would prove to have been a very bad investment for the Kremlin.

The Soviets' ties to Cuba had to be much stronger than those with any Middle Eastern client state. It had been a great thing in the beginning of the connection for the USSR to demonstrate the demise of the Monroe Doctrine and to have a pro-Soviet regime ninety miles from US shores. But here again there was a government over which—though it was Communist, unlike those in Egypt, Syria, etc.—Moscow did not have complete control. The memories of the Russians backing down in the 1962 crisis still rankled in Havana. Another reason for Castro's disenchantment with his allies was their failure, in his view, to support strongly enough his brand of "wars of liberation" throughout Latin America. The temperamental Cuban dictator thus required not only continuous economic help, a not negligible burden on the Soviet treasury, but repeated assurances that Moscow would not kowtow to Washington. In 1970 American intelligence discovered that the Soviets were constructing a nuclear submarine base in the port of Cienfuegos, something that ran clearly counter to the Kennedy-Khrushchev understanding of 1962. Washington sternly expostulated with Moscow. The Soviets promised to desist, but their submarines kept coming back to the port. Rather atypically, the State Department kept on protesting just as insistently, and the base never became operational.[9] By the middle of the decade, the USSR was to find another way of earning returns on its Cuban investment, and Castro that wider scene of activity which he craved so much and which had made him throw in his lot with the Russians in the first place.

Many of the Soviet world-wide commitments with their troublesome implications were a legacy from the Khrushchev years and bore no clear relationship to the national security, nor, assuming these were still of importance, to the ideological interests and self-proclaimed mission of the Soviet state. But much as Brezhnev and Co. may have decried their predecessor's "hare-brained schemes" and regretted this overextension of Soviet power and resources, they would not enter upon a policy of retrenchment. There was an element of deliberation as well as what one might call political snobbery in the

9. Kissinger, *op. cit.*, p. 651.

Soviet Union thrusting its influence into every corner of the globe. It kept the West on the defensive: troubles in the Middle East, Africa, Indochina distracted the American and Western European policy-makers and did not leave the US and its allies much leisure and resources to pressure the USSR closer to home, on Germany and Eastern Europe. The Soviet Union was now a great power, and thus like Great Britain in the nineteenth century and the US after 1945, entitled to have its presence felt and voice heard in every part of the world. Hence also the rapid build-up of the Soviet fleet, which no longer, as under Stalin, would have mainly a defensive function and consist mainly of submarines, but would challenge that of the US for mastery of the oceans.

The great-power motif was used occasionally, and rather disarmingly, by Soviet propaganda in trying to persuade the West that it should not become unduly disturbed by what the Soviets were doing, say, in southern Africa or the Caribbean. Moscow, the argument ran, did not really want to be a catalyst of every conceivable kind of crisis throughout the Third World. But it was incumbent upon it to make its presence felt everywhere. In a sense the Russians had been underprivileged for well-nigh two centuries by being denied that world-wide imperial role which their national greatness entitled them to play, and so one had to understand and should not be alarmed by their present self-indulgence on that count. All they really wanted was to be recognized as an equal of the US, and did not the latter have bases, economic interests, etc., all over the world?

This touching theme clashed sharply with that heard when Moscow addressed the Third World. Here the USSR was presented as the main force for liberation, protecting the new nations from the neo-imperialism of their former colonial masters and the US, their friend and protector. Russia's history even before the Revolution, this audience was assured, had always been free from the taint of colonialism and its related evils, racism and economic exploitation. Hence what the Soviets were doing in Africa and Asia could in no sense be likened to the past or present practices of the West. Theirs was,

on the contrary, "selfless help" for the cause of colonial emancipation, international justice, and peace. True, the credibility
of this argument has suffered somewhat as a consequence of
the Sino-Soviet dispute. Unlike the West in its current mood
of contrition for its real and alleged sins of the past, the Chinese
had no inhibitions in branding the Kremlin as the most per-
nicious imperialist force of all, and in pointing out that for all
of their alleged abjuration of colonialism, the Russians have
somehow managed to occupy one-sixth of the earth's landed
surface. But unlike Beijing, Moscow was capable of furnishing
massive military and economic help to the emerging nations.
It did so regardless of the given country's ideological colora-
tion, provided such help would serve some Soviet, usually anti-
Western, purpose. A few Third World countries have indeed
developed considerable skill at extracting economic assistance
from both sides while being beholden to neither. But most new
regimes were too unstable to be able to afford such manoeuvres, and felt they had to make a choice. The Americans
to be sure were likely to be more generous, but by the same
token more squeamish: there would be Congressional delegations, intermittent admonitions that the local regime follow
the straight and narrow path toward democracy and economic
development. The USSR usually took a more indulgent view
of the nondemocratic ways of various ruling elites: political
persecution or even outright terror as practiced, say, in Idi
Amin's Uganda, did not disturb the Soviets unduly, provided
the government indulging in them was "progressive," i.e., if
it adopted an anti-West stand or showed promise of moving
in that direction.

Another legacy of the Khrushchev era (though precedents
for it go back to Lenin's time) was the Kremlin's doctrine of
"wars of liberation." In practice it meant that Moscow had
arrogated to itself the right to determine which insurrection-
ary forces struggling against their colonial masters, or for that
matter against their own country's regime, were genuinely
fighting for independence and/or social justice, and which
represented an unjustified rebellion against a legitimate au-

thority. The criteria according to which the Kremlin formulated such judgments do not have to be spelled out.

In trying to ride the wave of anticolonialism, the USSR again enjoyed certain obvious advantages over the United States. The latter, for all of its professions of sympathy for all nations struggling for independence, had ties of long standing with the old colonial powers. No one was much surprised or shocked anymore when the USSR, while maintaining formal diplomatic, etc., relations with a given country, at the same time more or less openly encouraged subversive activities against its regime. But for the United States such double bookkeeping in its foreign relations did not come easily. It had to view sympathetically the efforts to rid Africa of the relics of European colonialism, and hence the CIA maintained direct contacts with the African rebels fighting Portuguese rule in Angola. Apart from traditional American anticolonial sentiment, such operations were viewed as necessary, if credit for helping the rebel cause, bound ultimately to prevail, was not to go exclusively to the Soviet Union. At the same time Portugal was an ally of the US, a fellow member of NATO. And very soon stories of such double deals would surface in the American media and become the subject of spirited Congressional debates as to whether and how far it was permissible for a democracy to indulge in such hanky-panky.

The global character of Soviet-American rivalry posed obvious difficulties for a *rapprochement* of the two powers. Insofar as the Third World dimension of that rivalry was concerned, Moscow was, as we have already seen, in an almost unassailable position in rebuffing American criticisms or even proposals to negotiate the problem: the Americans themselves had endorsed the principle of emancipation of colonial and oppressed people. How could they then object to the Soviets practicing what they themselves preached and occasionally (but, oh, how clumsily) took a hand in? Washington and other Western capitals, Moscow's argument continued, were obviously unhappy about Soviet sponsorship of "liberation movements" in regions like southern Africa and the Arabian

Peninsula, close to the sources of raw materials crucial to the West. But didn't the US repeatedly denounce the very notion of spheres of influence? How illogical then for America to imply that the USSR had no business interfering in the affairs of such areas and they should be left to themselves, in fact to the American neo-imperialists!

Behind the Marxist-Leninist idiom in which the language of Soviet diplomacy was couched, it operated very much in the manner of old-fashioned imperialism, with its penchant for secret agreements, bargains struck with another power at the expense of other parties, and spheres of influence. Such practices clashed with the moralistic and legalistic strain of the American approach to foreign relations, and quite apart from other considerations, made negotiations, and even the process of communication between the two superpowers, very difficult.

The President and his chief foreign policy adviser were perhaps of a more pragmatic turn of mind than any of their predecessors since World War II. But as already noted, they could not escape the constraints which the American system places upon the executive when it comes to foreign relations. Nixon hoped to use the SALT talks as a means to help him secure a favorable settlement of the Vietnam war, one which would enable the US to disentangle itself from Indochina without turning over the area to the Communists. Thus (to anticipate our story) before his visit to Moscow early in 1972, Kissinger had been instructed to warn the Soviet leaders that no strategic arms agreement was possible as long as the war was still going on.[10] Kissinger was also to point out to Brezhnev that the famous doctrine associated with his name was really incompatible with Soviet support of wars of liberation. Indeed the logic was impeccable: why should the USSR have the right to intervene whenever it saw fit in the internal affairs of Czechoslovakia or Hungary, while America's actions to protect its protégés and allies were branded as immoral and imperialist? But the Soviets were by now sophisticated enough about American politics to know that Washington, having committed itself

10. Kissinger, op. cit., p. 1136.

strongly to detente, could not pull back from SALT its central feature, without sorely disappointing American public opinion, especially in a presidential election year. For its own good reasons, Moscow would find it advisable to help arrange a truce in Vietnam, but not in a way that could suggest it was responding to American pressure or sacrificing the interests of Hanoi.

Officially the Soviets resisted and denounced any attempts at "linkage," as this technique of bartering one agreement for another became known (the wide usage of the term was probably due to the fact that though fairly synonymous with power politics, "linkage" to the American ear somehow did not carry the same sinful connotations). Every issue of international politics, Moscow staunchly asserted, ought to be discussed and settled on its own merits, and not through undignified horse trading. Nor was the USSR to be treated like an unruly child, with agreements conditioned by what the Americans arbitrarily assumed should be its proper behavior: no SALT unless you help us in Southeast Asia, no credits and trade unless you let Soviet Jews emigrate!

In fact the Soviets, when it came to something they badly wanted, have themselves been strong believers in linkage, but practiced this art with considerable subtlety. It finally dawned even on the Americans that one could not negotiate with the Russians for very long without a sudden surfacing of an issue seemingly unrelated to the one under discussion. The SALT negotiators were relaxing from their labors by attending a Rostropovich concert when the chief Soviet delegate passed a note to his American counterpart. No, it was not a comment upon the music. The note proposed "a bilateral agreement providing for joint retaliation against a country that launched a provocative nuclear attack. . . . Both sides would obligate themselves to take a retaliatory action if a third country committed such aggression."[11] The American negotiators and their Washington superiors became first puzzled and then apprehensive: what could the Soviets have in mind? Was it perhaps China? Ambassador Semenov, when questioned on the sub-

11. Smith, op. cit., p. 141.

ject, explained (undoubtedly with considerable inner exasperation) that no: "The agreement would be universal, not aimed at any particular country . . . if unreasonable leaders come to power in Country X and plotted a provocative attack, it would apply to that country." [12] The American chief negotiator, Gerard Smith, had finally the illumination that the Russians "likely saw some possibility that in a strategic arms limitation environment, they could float a nuclear superpower design to achieve global political aims going beyond arms limitation." And he concludes, a trifle naively, "They gave the appearance of having an emotional commitment to this notion which though Russian, seemed un-Soviet like to me." [13] It would have been useful, one would think, for the Americans, without necessarily committing themselves on anything, to draw out the specifics of this intriguing idea. Certainly Talleyrand or Metternich would have said: "How interesting, tell us more." But Washington would have none of that: Beijing or worse, some US senator, might hear about it. Though a veteran of the Stalin school of diplomacy, Ambassador Semenov allowed himself what this time really was an un-Soviet-like show of emotion: by refusing to discuss the contingency of a provocative attack by a third party, the US had passed up a golden opportunity, he declared, quoting a line of Pushkin's, "And happiness, before it glided away, was so near." [14] The Soviets, as we have seen, were genuinely, and not unreasonably, concerned about the possibility of another power triggering a nuclear confrontation between the US and the USSR. But the Soviet diplomat's poetic outburst was yet another piece of evidence of how under the cover of SALT negotiations, the Kremlin sought to inveigle the US into some kind of collusion against China.

Detente then would of necessity have to be limited in scope. What Washington considered the Soviets' destabilizing activities in various parts of the world could not be meaningfully discussed; at most they could be glossed over by pious for-

12. *Ibid.*, p. 142.
13. *Ibid.*, p. 143.
14. *Ibid.*

mulas that might or might not have any tangible consequences. The Chinese angle, so important in turn to Moscow,
also could not be directly addressed. It would be addressed
only obliquely.

From the Soviet point of view, a *rapprochement* with the
US was still expected to produce solid benefits. One might
well imagine Brezhnev putting the case for detente before the
Politburo in terms like these: "We need detente to avoid dangers inherent in a situation where the US, smarting from its
Vietnam humiliation, might be tempted to recoup its prestige
by drastic actions elsewhere, and where China, having shaken
off the paralyzing effects of the Cultural Revolution, might seek
an outright alliance with the West. More amicable relations
with America need not put undue restraints on our policies.
The actual scope of detente will depend largely on the evolv-
ing pattern of the international situation, which does not favor
the West. If, as is likely, the latter continues to decline in
strength and cohesion, detente should enable us to prevent
this decline from being accompanied by violent convulsions
that might set off a nuclear conflict. If the West recovers from
its faltering course and there is a new spirit of realism and
resolution in US policies, then under the umbrella of detente
we could strike some mutually profitable deals. To continue
the old way of dealing with the US, i.e., to try merely to outshout and outscare the Americans, would profit no one but
Beijing."

Some Soviet leaders must have been concerned about the
secondary effects of detente. Official propaganda would have
to soft-pedal the theme of the "imperialist West" constantly
plotting against the socialist camp: the theme that provided
much of the rationale for the repressive domestic practices in
the USSR and its satellites. Once the Americans ceased to be
so much afraid of what the Soviet Union might do to them,
they would inevitably develop an unhealthy interest in what
the regime was doing to its own allies and citizens. Dissent
would have to be dealt with, if not more leniently, then more
decorously; there might be pressures from Washington on
Moscow to alter its policies concerning Jewish emigration.

Greater exposure to the West had always brought with it the risks of ideological pollution carried by the hosts of foreign tourists and by Western journalists probing into things that were none of their business.

But the Politburo's consensus had to be that while there were obvious risks and costs in less strained relations with the US, those were more than outweighed by the expected benefits. Besides achieving some objectives it had aspired to since 1945, the USSR would profit by more extensive trade with the West and easier access to its advanced technology. And the latter was especially important in view of the declining rate of growth of the Soviet economy.

We need not ascribe the pursuit of detente to some Machiavellian schemes on the part of the Kremlin, in which the very concept was an opiate to dull the West's vigilance. If the capitalists gave way to the delusion that detente meant some fundamental change in Soviet foreign policy, it would be their own fault.

The Soviets made no secret of their expectations that one of the results of detente should be a certain loosening of the ties between the US and its Western European allies. By subscribing to SALT, the US was acknowledging that the USSR was its equal in strategic weapons; hence the value of the American "nuclear umbrella" over the West was somewhat depreciated. At the same time the broader aspects of detente were conducive to the belief that Moscow no longer, if ever, harbored aggressive designs against the West. The individual Western European states, the Kremlin thus calculated, should become less nervous about the Soviets' intentions, while at the same time growing more impressed with Soviet power and skeptical about the US ability to guarantee their security. The end result of such disparate impressions would be the Western Europeans' willingness to negotiate directly with Moscow, rather than, as hitherto, let the US speak on their joint behalf. The Americans could not be overly happy with such by-products of detente, but the Soviets might well have pointed out that they were practicing what a US president had once advised: "Speak softly and carry a big stick."

Even before it was formally sealed by President Nixon's visit to Moscow in 1972, detente began to bring the Soviets solid benefits. One of the most tenaciously held goals of the Kremlin's diplomacy had been to normalize the territorial status quo in east central Europe, i.e., to obtain formal recognition by the US and its allies of East Germany and of Poland's western borders. Up to the mid-1960s Washington, and especially Bonn, would not even hear of giving an official seal of approval to the division of Germany. Thwarted repeatedly in exacting these concessions directly, Moscow for some years had been promoting the idea of a general security conference which by proclaiming the inviolability of all existing European borders could incidentally achieve this goal. But the US and the Federal Republic did not budge.

And now it was Bonn that took the initiative in meeting the Soviets more than halfway. Chancellor Brandt's socialist government, which came to power in September 1969, was determined to change the hitherto prevailing pattern of relations with the Communist bloc. In pursuance of this new *Ostpolitik* the Federal Republic announced in November 1969 its adherence to the nuclear nonproliferation treaty and on August 12, 1970, signed a formal agreement with Moscow acknowledging the inviolability of the territorial status in Central Europe—that is, recognizing in effect the East German regime and Poland's right to those formerly German territories that had been placed "provisionally" under its administration by the Potsdam Conference. This recognition by the Federal Republic of what it had, with US support, tenaciously refused to accept ever since its birth in 1949 was further underlined by formal treaties with Poland and finally, in December 1972, with Communist Germany, wherein both sides agreed to exchange representatives and to place their relations on a new cooperative footing. Though there was no German peace treaty in the formal sense, the sum total of the above agreements and of America's concurrence in them gave the Russians practically everything they had sought to attain by such a treaty ever since the war.

From one point of view, the West had conceded only what

had been facts of international life for more than twenty years, and thus removed a major cause of intermittent friction and confrontation. From another, in psycho-political terms, the Soviets had scored a brilliant diplomatic success. The US and its allies acknowledged the legitimacy of a regime they had repeatedly denounced as based on brute force, a regime that in order to stop mass flights of its citizens to the free world (over three million between 1945 and 1961) had had to resort to the infamous Berlin Wall. Brezhnev and his colleagues could congratulate themselves on accomplishing through patient diplomacy what neither Stalin, with his blockade of Berlin, nor Khrushchev, for all his stratagems and threats, was able to achieve.

The only concession the Western powers were able to extract in return for acknowledging the division of Germany was the Soviet pledge not to interfere with the status of and free access to West Berlin. That also was the subject of tortuous negotiations between the Four Powers that in 1945 had assumed the occupation of Germany, negotiations which lasted from March 1970 to September 1971, with the final agreement ratified in June 1972 after the Soviet-American summit meeting. A cynic might have commented that the West had surrendered an important moral principle in return for the Soviets' promise not to indulge in the kind of practices they had no right to resort to in the first place. Well, in any case, fears of a possible new conflict, tension, and near-confrontation of the kind produced by the Berlin crisis of 1948–49 and the intermittent one of the years 1958–62 were now allayed, even if, in view of the previous Soviet record on Germany, they could not be entirely dispelled by the agreement. The USSR assumed the responsibility for preserving unimpeded Western access to Berlin across the intervening East German territory. This, unwittingly, provided an ironic commentary on the Soviet claim that—to give it the official name—the German Democratic Republic was a sovereign state. But even on Berlin, the tenacious Soviet diplomacy managed to extract a few ounces of flesh: the US and its allies had to agree that West Berlin was not politically a part of the Federal Republic and that the

Bonn authorities would refrain from demonstrative gestures such as holding meetings of their parliament there.

Germany had been the main battleground of the Cold War. In 1972 it could then be held as a good augury for detente that there, at least, the East and the West were able to settle their difficulties. In addition to the diplomatic coup it scored, the Kremlin had other reasons for satisfaction. The agreement cleared the way for greatly expanded trade with the Federal Republic, now an industrial giant, and opened up expanded access to its technology (in some fields as advanced as that of the US), all under favorable credit conditions. Similar benefits would flow to other Communist countries of Eastern Europe.

In allowing and encouraging their protégés to trade extensively with and to borrow from the West, Brezhnev and Kosygin might well have congratulated themselves on being more clever than their old boss, who in 1947 had spurned the Marshall Plan and forced Poland and Czechoslovakia also to reject capitalist largesse. Both of those countries had recently required massive infusions of Soviet economic assistance, Czechoslovakia after the invasion in 1968, Poland following the workers' riots that toppled its Communist leaders in December 1970. Now, hopefully, the capitalists were going to relieve the Soviet economy of such burdens and thus help maintain the stability of the Communist regimes. At the time, and throughout the seventies, such expectations seemed amply justified, and it was only the Polish crisis of 1980 (precipitated and aggravated by the government's unwise use of massive Western credit) that showed that Stalin may not have been wrong after all in urging Communists to beware of capitalists bearing gifts and loans.

What made other features of detente possible was of course the strategic arms agreement. As we have seen, the USSR was frustrated in its efforts to impart a broader political (mostly anti-Chinese) dimension to the SALT talks themselves. But otherwise, the end result of Soviet-American negotiation as it crystallized toward the end of 1971 promised to meet most of the Soviets' objectives. This is not to say that, as many American politicians and military experts claimed subsequently, the agreements signed in May 1972 represented a one-sided Soviet victory. The USSR had made some concessions, from its point of view considerable. And, as on many previous occasions, it was not so much the Russians' negotiating skill and tenacity as the Americans' previous errors of omission and commission that in perspective make us see SALT I as having benefited the Soviets more than the US.

The end result of SALT reasserted, even if by implication, Soviet military superiority in Europe vis-à-vis NATO. The Kremlin also sought and obtained acceptance of its quantitative superiority in several categories of nuclear arms. Both results had been largely predetermined by US military policies of several years standing, and by the current political situation in America.

The formal acknowledgment of the "essential strategic equivalence" of the two powers was bound to have a profound psychological effect on America's European allies. From the inception of NATO, it had been taken as almost axiomatic that the USSR would always hold an edge over the Western allies insofar as military manpower and conventional arms were concerned. The ultimate deterrent to a Soviet attack on and takeover of Western Europe lay in America's strategic, i.e., nuclear arms, *superiority*. Now, by Washington's own admission, that superiority had disappeared and presumably could not be recouped. All that the US now claimed (and in a few years even this would be questioned) was that it had enough strategic power to deter any possible Soviet nuclear strike upon *its own territory*. But, as some in Paris, Rome, and Brussels were bound to wonder, what if the Soviets attacked and overwhelmed the West with conventional arms? Would an American President then bring himself to authorize an American first strike on the USSR, fully realizing that a Soviet nuclear response would lay waste his own country and bring death to untold millions of Americans? Probabilities were still against the Soviets taking such horrendous risks, but what had looked fairly reassuring in the 1950s was bound to seem much less so after 1972.

The Soviets, as we have noted, were well and happily aware

of the further strains SALT would place upon NATO and the Europeans' confidence that America would protect them if . . . But as someone in the Kremlin might have reflected, the responsibility for such developments lay squarely upon the US and its allies, rather than on some nefarious stratagems and ruses by the USSR. When NATO was being formed in 1949 its European members, still recovering from the effects of the war, naturally felt incapable of matching the Soviets in Europe man for man and tank for tank, and consequently NATO's forces were designed to delay a hypothetical Soviet advance until the US could bring into play its overwhelming air-nuclear superiority, rather than to throw back the Russians. But by the mid-1960s at the latest, the NATO strategy should have been changed. NATO had been considerably enlarged, especially by the accession of West Germany. Western Europe had gone through several years of prodigious economic growth. Its gross national product was considerably bigger than that of the USSR, and it surpassed the latter in population. There was no reason why the assumption of NATO's conventional arms inferiority, reasonable in 1949, should still have persisted in 1965. It was then fully within the capabilities of America's European allies more than to match the Warsaw Pact's forces; yet they continued to profess (though ever less confidently) their reliance on the US's now increasingly porous nuclear umbrella to save them both from the Russians and from having to spend more on defense. Had Washington been more mindful of General de Gaulle's sensitivities, had it chosen the right moment to suggest that its military presence in Europe might be reduced unless the allies' own contribution to common defense was increased, the negative effects of SALT I on the general balance of power would have been mitigated. But Washington's preoccupation with Vietnam and de Gaulle's real and imagined grievances against the "Anglo-Saxons" prevented the West from seizing the right moment for an appropriate restructuring of NATO, which—quite apart from its military advantages—would have been conducive to greater Western unity on political and economic issues as well.

What must have most impressed an average newspaper

reader trying to decipher the actual texts of SALT I was the quantitative superiority they conferred upon the USSR. The final agreement froze the current (1972) number of fixed ICBM launchers at 1618 for the USSR and 1054 for the US. The Soviet Union was permitted to build up to 62 modern nuclear-armed submarines, the US only up to 44. Of the three main categories of nuclear delivery systems, America retained an edge only in heavy bombers (not covered explicitly by treaty), 460 against 140. Our layman had additional reason to conclude that the final form of the strategic arms agreement favored the Soviet side, upon learning that for the most part its missiles carried heavier megatonnage and hence possessed greater destructive capacity than those in the US armory. Russia then seemed to have gotten the best of the bargain and for the next five years at least (the duration of the agreement "on the limitation of Strategic Offensive Arms") was bound to have an edge on the US insofar as strategic weapons were concerned.

Experts on the American side hastened to reassure the public: the numbers were far from telling the whole story. The US was considerably ahead in nuclear technology. Its land-based missiles, though smaller, had much greater accuracy than the Soviet ones, its submarines were better constructed and harder for the enemy to locate. Above all the US was way in advance of the Russians when it came to the development of MIRV, a single missile equipped with several independently targeted warheads. Such arguments, for all their undoubted validity, could not erase the impact of the statistics, which had to lead to some uncomfortable reflections on the part of both statesmen and voters of the non-Communist world. And even the expert had to remember that when it came to weapon technology, American advantage over the Soviets has seldom been long lasting.

American acquiescence in quantitative inferiority was again a legacy of a long-standing neglect of the psycho-political aspects of the arms problem. In purely military terms it possibly made sense for the Pentagon to opt for a smaller but more accurate missile, and to eschew, beyond a certain point, the expensive and apparently pointless production of weapons and delivery systems. But this reflected the tendency "to envisage nuclear strategy in the guise of warfare, rather than as an extension of international politics." [15]

American apologists for the 1972 agreements were subsequently to claim, and quite justifiably, that they had had few bargaining chips left in their negotiations over SALT I: "As a result of decisions of our predecessors, no American programs existed which could possibly produce new missiles for at least five years," wrote the main architect of US foreign policy. [16] One remaining bargaining chip had been America's superior technology when it came to constructing antiballistic missile systems—ABMs. But many politicians and experts alike believed that such systems, while conceivably effective against a limited nuclear attack launched, say, by China, would be largely useless against the full panoply of strategic weapons possessed by the USSR. Furthermore, the Congress, in its current mood of irritation with defense policies brought about by the experience of Vietnam, was not likely to authorize huge expenditures for an extensive ABM system, even if the Russians went ahead with their own antiballistic missile devices. And finally, both sides agreed that each would be limited to two ABM systems: one to protect its capital, another an offensive missile installation, neither to be equipped with more than one hundred interceptor rockets.

For the moment—and this was to last until the mid-1970s when the whole framework of detente began to crumble—American public opinion tended to be impatient with the broader implications of nuclear arms policy. Hence the negotiations with the Soviets were viewed with a mixture of fatalism and hope. Once it came to an actual war all bets would be off, so what was the use of worrying about numbers and megatonnage? Then too, the mere fact of negotiations and agreement with the Russians had to lead to restrained but real satisfaction. It gave the promise of human control of and ap-

15. Edward Luttwak, *Strategic Power, Military Capabilities and Political Utility* (Beverly Hills, Calif., 1967), p. 10.

16. Kissinger, *op. cit.*, p. 821.

plying rational criteria to something which had seemed both uncontrollable and too horrendous to be subjected to quotas and regulations. Soviet compliance with the agreement, the public was assured, would not depend merely on the Russians' word. The problem of verification, hitherto one of the main stumbling blocks to any arms limitation treaty, was now solved by the use of spy satellites, with both parties agreeing not to interfere with the other's monitoring of the number of its launchers, etc.

Most important for the wider public, was what seemed to be the Soviets' implicit acknowledgment that a nuclear war was too terrible to be a viable instrument of national policy.

Only if one remembered those sudden and drastic changes in America's mood concerning foreign affairs could one anticipate that in a few years many of these attitudes would be drastically reversed. SALT I, a great achievement in 1972, by the end of the decade would be seen by many to have been a trap. Those quantitative elements of the strategic arms picture that were undervalued in the 1960s would by the late 1970s assume an exaggerated importance in the minds of some specialists and politicians. There would grow a suspicion that the Soviets had never accepted the doctrine of mutual deterrence, and did not foreclose the possibility that by striking first they could score a meaningful victory in a nuclear war.

Before the final bargain-agreement could be sealed, the two sides engaged in a number of manoeuvres that were to define the limits of detente. Both Washington and Beijing had for some time been eager to effect a *rapprochement* prior to any Soviet-American summit meeting. But it suited the Chinese purpose to play the part of the wooed rather than the wooer. The Nixon Administration for its part did not object to the latter role, since it led to the impression that it was mainly the President's bold diplomacy that overcame Communist China's hitherto implacable enmity toward the US. For all the deep secrecy in which Kissinger's July 1971 trip to Beijing had been veiled, Moscow could hardly have been surprised by the revelation of it when his hosts extended an invitation to the President. Still the Russians are not invariably good at concealing

their nervousness. After the announcement, Ambassador Dobrynin inquired whether Nixon would not care to come first
to Moscow before visiting the People's Republic.[17]

The Soviets soon had an opportunity to remind the world
that for all its enormous bulk and potential, China was still far
from being a great power. Developments in the Indian Peninsula raised once more the prospect of war between its two
states. Truncated Pakistan was breaking up, East Bengal having for some time been in revolt against the central authorities. India's perennial hostility to its neighbor had hitherto been
restrained by the fear of China, Pakistan having been for years
Beijing's only non-Communist friend in Asia. Until 1971 the
Soviets had been urging moderation upon New Delhi; in 1965
they had hastened to arrange a truce between India and Pakistan, obviously fearful that a Chinese military intervention
could precipitate an uncontrollable chain of events. Now, on
the contrary, Moscow encouraged India to step up its assistance to the East Bengal rebels and thus to offer open defiance
not only to Pakistan, but also to the Chinese Communists. Pakistani authorities had acted as intermediaries between Washington and Beijing. And the Russians had good reason to believe that this time the latter was in no position to intervene
militarily on behalf of its Pakistani friends. Communist China
had been weakened by the excesses of the Cultural Revolution. With Mao's reign obviously nearing its end, the leadership of the country was deeply divided by factional strife, vividly illuminated by the mysterious death and subsequent
official denunciation of Marshal Lin Biao, until recently the
designated heir to the aged Chairman.

On August 9 the USSR and India signed a treaty of friendship which provided for consultation and help should either
of the signatories be threatened by a third party. There followed a flurry of high level Indian-Soviet consultations as India's military confrontation with Pakistan grew more intense,
by November turning into open warfare. China's reaction remained confined to threats. In the UN Security Council, the
Soviet Union blocked a US-sponsored resolution calling for a

17. Kissinger, op. cit., p. 766.

ceasefire and Indian troop withdrawal from East Pakistan. There was a last-minute, rather clumsy attempt by Washington to warn off New Delhi—an American task force headed by an aircraft carrier lumbered into the Bay of Bengal. By then, the middle of December, Pakistan's armies in the east had already been crushed. East Pakistan became independent Bangladesh. China's weakness and the value of the Soviet Union's support had been vividly demonstrated. It was characteristic that soon after their bitter denunciation of the USSR for its complicity in their country's partition, Pakistan's leaders sought to normalize their relations with the Soviets. The country's new president, Zulfikar Bhutto, paid a state visit to Moscow in March 1972. Insofar as India was concerned, Indira Gandhi's regime (and she has remained her country's dominant political figure to this day) would continue on its friendly course toward the Soviet Union and remain censorious vis-à-vis the United States.

For all of China's temporary weakness and the confusion in its councils, the Soviets could not delude themselves that militarily or otherwise their alliance with India could in the long run dispose of the Chinese danger. The Kremlin's policy then followed a two-fold course. One, as spelled out by Brezhnev at the Twenty-Fourth Party Congress in 1971, consisted in insisting that the USSR was always ready to resume its former ties with the People's Republic, and that if the latter wished sincerely to reestablish comradely relations the Soviets would not be found wanting. The aged Great Helmsman's disappearance from the scene could not be much delayed (Mao had been born in 1893), and after his death, it was hoped, one at least of the contending factions might seek support from Moscow. Along with such rather slender hopes went unremitting efforts to isolate China within the world Communist movement and to contain it in Asia. Neither of these appeared to be succeeding. Most foreign Communists, no matter how close to Moscow, would not agree to a formal condemnation of Beijing and placing it beyond the pale. The idea of an Asian security treaty, intermittently advanced by the Brezhnev regime, also found few takers among the continent's states. They all under-

stood its anti-Chinese implications, and most of them preferred to strive for good relations with both giants, rather than
to incur the wrath of either.

President Nixon's visit to China took place in February 1972.
The historic nature of the occasion obscured for the Americans, but probably not for the Chinese or the Russians, its inherent irony. America's massive involvement in Vietnam had
been largely rationalized in terms of its potential effects on
China. In the early 1960s the latter was believed, and justifiably, to be a much more militant propagator of Communism
than the USSR. Defeat of the Vietcong was to teach Beijing a
lesson: the wars of "national liberation" it instigated could
not prevail when America's power was thrown onto the scales.
Now, largely as a result of the war and America's discomfiture
on its account, the US had de facto, if not yet formally, recognized the Communist regime and agreed to its replacing
Taiwan in the United Nations; and the President was seeking
his hosts' help to extricate his country from the war in Southeast Asia. There was to be sure an even more bitter disappointment in store for the Chinese. It was they who had urged
North Vietnam toward a more aggressive course of action
against the South, while the Soviets had pleaded that Hanoi
remain patient and circumspect. But already now the Chinese
influence with the Vietnamese Communists as against that of
the USSR was declining. The Soviets had been providing the
North with vital war materiel, the Chinese limiting themselves
mostly to exhortations. And in a few years the Chinese leadership would have concrete reasons for second thoughts about
the outcome of the war it had helped start in Indochina.

The Vietnamese scene flared up between Nixon's Beijing and
Moscow visits. Hanoi had an obvious interest in escalating the
war before the US could seal any deal at its expense. More
northern troops poured across what was still, ironically, called
the demilitarized zone between the two Vietnams. The US responded with greatly intensified bombings. In April, B-52s
raided targets in the Hanoi-Haiphong areas, incidentally hitting four Soviet ships in the harbor. Simultaneously Nixon
dispatched Kissinger to Moscow to impress upon Brezhnev

and his colleagues that the summit meeting and all the Soviet-American agreements in the offing—SALT, trade, etc.—might have to be canceled unless the Soviets restrained their North Vietnamese protégés.

The President, as Kissinger per his instructions informed Brezhnev, was not a man to be trifled with: he was "direct, honest, strong, fatalistic—to him the [presidential] election is not the key. [He] will not be affected one iota by public opinion." [18] The General Secretary, for his own part, pleaded that he could not order Hanoi around, but only act as an honest broker between the two parties. As usual in such situations, the Soviets were careful to leave the door a little bit ajar. The President's assistant was encouraged to continue his confidential Paris negotiations with North Vietnam's emissaries. As the top man, Brezhnev could not give the impression of being less "direct, honest, strong and fatalistic" than his American counterpart, but lesser Soviet lights are occasionally permitted to appear more flexible and hopeful of softening their leader's iron resolve. And so Foreign Minister Gromyko hinted that the Russians were eager to help find a solution to the war. Thus encouraged, Kissinger discussed several items other than Vietnam pertaining to the coming summit (earning himself upon his return a stern reprimand from his boss), which in turn probably persuaded Moscow that it would not have to lean too hard on Hanoi in order to save detente.

Within a few days increased military activities by the Vietnamese Communists led the President to an outright test of wills with Moscow. He ordered the bombing and mining of the sea approaches to North Vietnam, the main artery through which Soviet supplies were being expedited to the country. Many in the West held their breath; surely the least Moscow could do was to cancel the summit. There was to be sure an outburst of Soviet rhetoric usual for such occasions ("unless . . . , the gravest possible consequences will follow"). But the invitation to the President was not withdrawn.

And so in May Nixon went to the Soviet Union, the first American president to do so in peacetime. Unlike previous

18. Kissinger, *op. cit.*, p. 1136.

encounters between the chief American executive and the top Russian oligarchs, productive of nothing beyond those short-lived and delusive "spirits"—of Geneva in 1955, Camp David in 1959, and Glassboro in 1967—this one was signaled by important agreements and an understanding ushering in detente. No leader on either side could believe that the meeting resolved all the tortuous problems of Soviet-American coexistence, still less that as Nixon put it, waxing lyrical upon the signing of SALT I: "The historians of some future age will write of year 1972 . . . that this was the year when America helped to lead the world out of the lowlands of constant war to the high plateau of peace."[19] Still, what one could realistically hope at the time was that the Moscow summit marked a significant first step toward a new pattern of relations between the two superpowers: that while their basic rivalry for influence and power throughout the world continued, they would find it possible to cooperate on a number of important issues.

The impressive array of documents and declarations signed in Moscow obscured for the moment the fact that neither side obtained what it had originally hoped for from the summit: the US a definitive pledge of Soviet help in settling the Vietnam war, the USSR some kind of understanding concerning China. Brezhnev hinted ever so gently that the latter might be a price for the former. Would it help, he asked toward the end of the summit, if he dispatched his Politburo colleague Podgorny to Hanoi? And almost in the same breath he observed that both their countries "might usefully keep an eye on the nuclear aspirations of Beijing."[20] And his American hosts should know, Kosygin told them rather disingenuously, that in 1965 the Chinese had been ready to send their troops to fight on the side of North Vietnam. The Chinese, the Soviet leaders all agreed, were a bad lot, their foreign policy thoroughly immoral. The Americans could have scored some debating points by inquiring why it was more immoral to help fellow Communists with troops rather than with arms, as the Soviets had done, and whether it was by hosting the Ameri-

19. Smith, *op. cit.,* p. 451.
20. Kissinger, *op. cit.,* p. 1251.

can president that Mao's regime displayed conclusively its utter lack of principles. But such frankness would have spoiled the somewhat contrived aura of cordiality in which the summit was taking place. And much as he was already being accused of indulging in "imperial presidency," Nixon simply could not say: "Do what you will about China, provided you help me find an honorable exit from Vietnam."

Publicly both sides recognized their differences on Vietnam. The Soviet side affirmed "its solidarity with the just struggle of the peoples of Vietnam, Laos and Cambodia" and firmly supported "the [peace] proposals of the Democratic Republic of Vietnam" and "a complete and unequivocal withdrawal of the troops of the US and its allies from South Vietnam."[21] The US pleaded for "an internationally supervised Indo-China–wide cease fire," but significantly did not restate its hitherto official position that North Vietnam should also pull out its troops from the South.

The mechanics of the summit provided an amusing sidelight on the Soviets' negotiating methods. While adhering to the proverbial Slavic hospitality, Brezhnev and his colleagues put their American guests through a hectic round of activities. Nixon's repose would be interrupted by his being "kidnapped" for a hydrofoil ride, or by the General Secretary, a great automobile buff, who would himself take the wheel for a hair-raising drive. But even amidst such playful distractions, the Soviet leaders would suddenly press important political questions upon their somewhat shaken visitors, raising the suspicion that it was not merely sheer ebullience that was behind this behavior.

The summit led to the consummation of the SALT process. The long-drawn-out Vienna-Helsinki series of talks on nuclear arms had for some time been paralleled by what came to be known as back-channel negotiations: direct exchanges on the subject between Moscow and Washington, with—as in many other confidential aspects of American policy—Kissinger, rather than the Secretary of State, acting as spokesman. This

21. Richard P. Stebbins and Elaine R. Adam, *American Foreign Relations: 1972 Documents* (Washington, D.C., 1976), p. 73.

double-track procedure, with the official American delegation to the talks often uninformed as to what was transpiring in the back channel, was subsequently blamed, possibly justifiably, for weakening this country's bargaining position.[22] But granted the circumstances and Nixon's determination that SALT should be the main feature of the summit, it is difficult to see how its provisions could have been much altered.

The arms limitation pact signed in Moscow consisted of two parts. One was a treaty of unlimited duration, but subject to termination on six months' notice by either party, restricting both powers to two ABM systems, one to protect the nation's capital, the other an ICBM launcher area of its choice. By a subsequent (1974) protocol to the treaty, the number of permissible ABM systems was reduced to one for each power. Even so, the USSR never completed its defensive screen around Moscow, and the US scrapped its projected ABM system around an ICBM launcher site. The development of MIRV rendered the effectiveness of antiballistic missile defense more than ever questionable.

The heart of SALT I was the agreement "on certain measures in respect to the limitation of strategic offensive arms" stipulated to remain in force for five years. We have already noted the numerical quotas to which it restricted the two powers' land (ICBM) and submarine (SLBM) launchers and the overall quantitative superiority which accrued under them to the Soviet Union (the upper limit in both categories for the US 1710, for the USSR 2350). Subject to certain restrictions, "modernization and replacement of strategic offensive ballistic missiles covered by the Interim agreement" was permitted.[23] And here the American proponents of SALT I were in for an unpleasant surprise: they believed that Soviet superiority in the number of missiles and their throw weight was more than balanced by superior US technology. The Soviets, it was confidently expected, would not be able for several years to MIRV their missiles. "Instead, little more than a year after

22. Smith, *op. cit.*, pp. 466–68.
23. US Arms Control and Disarmament Agency, *Arms Control and Disarmament Agreements* (Washington, D.C., 1980), p. 151.

the signature [of SALT I] three different Russian MIRV systems had appeared, sufficiently developed for advanced testing."[24]

To be sure the Soviets could and did claim that the respective numbers of ICBMs and SLBMs were far from telling the whole story insofar as the nuclear weapons balance between the two superpowers was concerned. The agreements did not put any constraints on strategic bombers, in which the US held a considerable superiority over the USSR. SALT I did not deal at all with tactical, i.e., short- and intermediate-range nuclear missiles. Such weapons in the hands of NATO could thus be modified so as to be capable of reaching targets in European Russia. Britain and France, America's allies, disposed themselves of small but not negligible stocks of nuclear arms and delivery systems. Finally, China's strategic weapons arsenal was growing, and of course now it had to be of much greater concern to Moscow than to Washington.

To sum up: we may be sure that the Kremlin appreciated the psychological advantages and hence potential political benefits accruing to it from SALT I, but nothing entitles us to believe that it also viewed it as enlarging its *military* options or making the prospect of a nuclear war less horrifying for the USSR. The Russians were genuinely concerned by the possibility of a nuclear exchange between the two powers being triggered by an accident or through a third party's provocation, and a separate agreement provided for immediate and urgent consultations "if at any time relations between the parties or other countries appear to involve the risk of a nuclear conflict."[25]

To emphasize the importance of the agreements, they were signed by the political leaders of the two countries, and not, as tradition would have urged, by their foreign ministers. There were also reasons touching on domestic politics that dictated this rather irregular procedure (especially in Brezhnev's case since at the time he was head neither of the state, nor of the

24. Edward Luttwak, *The US-USSR Nuclear Weapons Balance* (Beverly Hills, Calif., 1974), p. 34.

25. *Arms Control and Disarmament Agreements*, p. 160.

government). With the presidential elections approaching, Nixon was eager to stress his personal role on this historic occasion. And for the General Secretary it was a vivid demonstration that his was the dominant voice not only in domestic, but also in foreign affairs, the latter until recently the province of Prime Minister Kosygin.

The summit was also marked by a multitude of other agreements and declarations. For the most part they were concerned with noncontroversial matters: strengthening of trade ties between the two countries, scientific and cultural exchange—treaties which had been previously agreed upon and ordinarily would have been signed without much fanfare in the respective capitals. Their inclusion on the agenda of the summit was intended to strengthen the impression that this was truly a turning point in the relations of the two countries, indeed in world affairs. Detente was extended to outer space: in 1975 an American spaceship was to link with one manned by Soviet cosmonauts. And in fact this spectacular demonstration of the two powers' collaboration did take place as planned, but, alas, by that time Soviet-American relations here on earth had already suffered a setback.

Another feature of the summit may strike us also as a theatricality on a par with the projected joint space shot: Nixon, speaking from the Kremlin, addressed the Soviet television audience. Yet symbolically this was not unimportant. Television, along with other Soviet media, have traditionally and insistently excoriated the American ruling establishment as imperialist and addicted to war-mongering. Here, and rather confusingly for the average Soviet citizen, his own government lent its official sanction to projecting quite a different image of the leader of world capitalism. Nixon, not invariably successful in his television appearances before his own countrymen, made this time a moving appeal for peace and friendship between the two nations.

The Soviets' eagerness to please extended not only to the President and his entourage. The Kremlin realized the strong emotions aroused in the US by the plight of those Soviet Jews who wanted to emigrate to Israel, or at any rate to leave the

USSR. To ignore this sentiment risked jeopardizing detente, or at least those elements of it which depended on Congressional legislation. Probably no other aspect of the *rapprochement* with the US was subject to as much debate and lively disagreement within the Politburo as the question of whether and to what extent to humor the US on the Jewish question. Traditionally the Soviet Union brooked no interference in its domestic affairs. American concern with the issue must have been the source of deep irritation to the Soviet oligarchs. Under Stalin, whose repressive domestic practices made the Brezhnev regime appear by comparison a model of liberalism, the US policy-makers would not have dreamed of intervening on behalf of Soviet Jews or other subjects of the Kremlin. Even to admit the possibility of a foreign power's interference in matters pertaining to the sovereign rights of the USSR was hurtful to national pride. More than that: to tolerate such interference was ideologically embarrassing; it meant acknowledging that a sizeable element of an ethnic group found life in the Fatherland of Socialism unbearable. The Soviets' Arab friends could not be enchanted by their ally permitting its Jews to leave, and thus indirectly contributing to the strengthening of the Jewish state. To appease the Americans by letting more Jews emigrate could set a most unfortunate precedent for the future. The Americans would be encouraged to keep sticking their noses into Soviet internal affairs: today the Jews, tomorrow political dissidents in general, other ethnic groups claiming discrimination, and God knows what else. Equally intolerable could be the domestic repercussions of the Kremlin's yielding to such pressures: how many would be emboldened to articulate their grievances if instead of an exemplary punishment they could expect to be allowed to leave?

It is a strong proof of the importance attached to detente by the Soviet leadership that it did not tell the Americans to mind their own business. Of course there could be no question of the Soviet Union granting an explicit right to emigrate to any of its citizens who wanted to do so. Still the Soviets did listen to their American guests' remonstrations on the subject and strongly implied that yes, more Jews would be allowed to emigrate.

In fact even prior to the summit Soviet emigration policies had undergone a change. Of some 80,000 Jewish petitioners for exit permits between 1968 and 1971, roughly 18,000 had their wish granted. Hardly an impressive proportion. But one must recall the story of those Soviet women who during the war had been married to Americans and Britons stationed in the USSR. After 1945 none of that handful were permitted to follow their husbands home. Most had recourse to divorce, and it was only in 1953 after Stalin's death that this ban, equally astounding in its pettiness as in its cruelty, was withdrawn.

The Brezhnev regime's "liberality" on emigration was not solely the result of its concern about Western public opinion. There were also cogent domestic reasons. The Kremlin's long-standing anti-Zionism had, since the Arab-Israeli conflict, occasionally shaded into outright anti-Semitism. Many Jews, even those thoroughly russified and loyal to the regime, were subjected to chicanery in their professional and even private lives, and felt unwanted in the country of their birth. With mass terror no longer a viable option, the regime had every incentive to find other ways to get rid of the most vocal element among the malcontents. Some within the Soviet oligarchy would undoubtedly have opted for a wholesale expulsion of what they viewed as an alien element. But to banish outright some two million Jews would, apart from the loss of valuable human resources, have meant the most explicit repudiation yet of what Marxism-Leninism was supposed to stand for. Embarrassing though the whole matter was, the Soviets still found it possible to make a virtue out of necessity: because of their consideration for their American friends' feelings, they would allow more Jews to emigrate! During what might be described as the honeymoon period of detente, 1972–73, some 65,000 were allowed to leave.[26] In subsequent years the numbers were to vary, reflecting the general fortunes of detente. To be sure

26. Colin Shindler, *Exit Visa* (London, 1978), p. 271.

there was to be nothing automatic about the process; many petitioners for exit permits would receive outright refusals, and even the successful ones would encounter delays and experience harassment before being allowed to leave the land to whose culture their community had so greatly contributed.

International agreements based upon the solid foundations of common interest and shared values seldom need to be accompanied by a grandiloquent statement of principles. It was perhaps not a good augury that the two sides thought fit to conclude the summit with a solemn statement reasserting their hopes for the future. As might be expected it included ritualistic phrases saying that differences in ideology and social systems would not be allowed to bar "the bilateral development of normal relations" and that "in the nuclear age there is no alternative to conducting mutual relations on the basis of peaceful coexistence."[27] There was an assertion which some may have felt did not quite correspond, to put it delicately, to the facts: "[The USA and the USSR] recognize the sovereign equality of all states." Even states bound by the closest ties of friendship would have found it difficult to live up to the high-minded pledge that "efforts to obtain unilateral advantages at the expense of the other, directly or indirectly" must be eschewed as inconsistent with the new relationship of the two powers.

And yet it would be overly cynical to see the document—"Basic Principles of Mutual Relations between the United States of America and the Union of Soviet Socialist Republics"—as simply an exercise in bathos and wishful thinking. For the first time in a generation the two superpowers were able to survey the whole realm of international relations, and to do so without resorting to mutual vituperations and accusations. In conjunction with the actual agreements concluded at the summit meeting, the declaration could thus be taken as a harbinger of more efficacious communication and a more civil dialogue between the two superpowers than had existed ever since their joint struggle and victory over Germany and Japan.

27. Stebbins and Adam, op. cit., p. 76.

3

In the Wake of Vietnam
and Watergate

"Important and many-sided as the achievements of 1972 were, they may be viewed as but the first steps on the path" (of Soviet American *rapprochement*), wrote George Arbatov, one of the Soviet Union's leading official experts on the US.[1] In addition to being an adviser on American relations to the Politburo, Arbatov has also served as an unofficial Soviet envoy to the American intellectual community. The article in question was written for an organ of the Central Committee and for a select Soviet audience, and thus may be taken as representing one element of the Kremlin's thinking about detente.

To be sure, accompanying this hopeful assessment of the prospects for the future was an important reservation: "There can be no question as to whether the struggle between the two systems will or will not continue. That struggle is historically unavoidable."[2] But detente should strip this struggle of its element of danger. Instead of "military confrontations, an armaments race, and dangerous clashes," it would take the form

1. "Soviet-American Relations," *The Communist*, no. 3 (February 1973), p. 110.
2. *Ibid.*, p. 111.

of an "ideological rivalry and competition in many spheres," while allowing also for a degree of cooperation, "with the disputes which cannot be avoided being settled through negotiations, and the competition in arms giving way to their limitation and (eventually) disarmament." It was up to the US to be done with the survivals of Cold War mentality still lingering in certain remnants of the American ruling class, but on the whole the prospect was encouraging. The average US citizen had realized the futility and inordinate costs of his country's past imperialist policies. "The decisive victory of Nixon in the recent [presidential] election had as one of its main causes the highly favorable reaction of the electorate to . . . the agreements with the USSR."[3]

It is unlikely that in their private discussions the Soviet policy-makers adhered *exactly* to this rather rosy vista of the future sketched by their adviser and propagandist. Nevertheless as of the beginning of 1973 the Politburo undoubtedly envisaged quite a different scenario of Soviet-American relations than the one that began to unfold in less than a year. The USSR did not propose to give up its expansionist policy, or, to use the article's euphemism for it, "the [ideological] struggle between the two systems," but expected that the US, having learned its lesson—i.e., having acknowledged Russia as its equal—would not be able to bar the further spread of Soviet influence throughout the world. At the same time Soviet foreign policy would have to be circumspect and avoid undue provocation of the US. The latter, the Kremlin believed, was on the point of recovering from its Vietnam malaise with Nixon firmly in the saddle (and the Soviets had, and continued to have even after the majority of his fellow citizens lost it, considerable respect for the President's political perspicacity and resolution). Also hope that the two countries could somehow arrive at an agreement about and common front vis-à-vis Beijing—a theme that had appeared intermittently in Soviet policies since the late 1950s—was still not dead.

By the same token, the idea of "the struggle between the two systems," i.e., of undermining US interests whenever and

3. *Ibid.*, p. 108.

wherever possible, had in the Soviet rulers' minds often and incongruously coexisted with a hankering after what the Chinese not inaccurately described as "co-hegemonism." In 1961, at the time of the greatest tension and danger of confrontation over Berlin, Foreign Minister Gromyko still could say, almost wistfully: "Our country places special importance on the character of the relations between the two giants—the Soviet Union and the United States. If those two countries united their efforts in the cause of peace, who would dare, and who would be in a position, to threaten peace? Nobody! There is no such power in the world."[4] At other times, and for reasons which have already been abundantly discussed here, the Kremlin realized the practical impossibility of the US and Russia entering into a partnership to rule the world. Yet there must have been occasions when the more thoughtful Soviet leaders reflected ruefully as to whether some new coup of theirs in Angola or South Yemen was really worth the risk. That cautionary motif and recognition of the inherent danger of imperialist policies in a nuclear age steals into the otherwise optimistic survey by Arbatov. "In a world where international tension is the rule, where there exist smoldering military conflicts in many places, states may be drawn quite unintentionally into a sequence of events in which eventually they lose control over the situation and it becomes impossible to prevent a catastrophe."[5]

The above is rather unusual; it is very seldom that one sees an authoritative public Soviet statement that as much as hints at the possibility of the USSR finding itself in a situation over which it has lost control and of war resulting not from some "imperialist aggression" but from a sequence of events whose final and catastrophic outcome was willed by neither side.

But as against such an atypical if veiled avowal of the risk inherent in its expansionist policies, the bulk of the Kremlin's official pronouncements continued to stress that the main lines of Soviet foreign policy would remain unchanged by detente.

4. *Twenty-second Congress of the Communist Party of the Soviet Union* [stenographic report] (Moscow, 1962), vol. 2, p. 343.

5. Arbatov, *op. cit.*, p. 109.

The Soviet-US agreement quite naturally led to some agitation among those governments and movements that were beneficiaries of Russian help and were now anxious lest detente lead Moscow to abandon or reduce substantially its backing of the "wars of national liberation" and other anti-Western activities it had traditionally supported. It was only a few weeks after the Nixon-Brezhnev summit, on July 18, 1972, that Sadat ordered 20,000 Soviet military personnel out of Egypt. Publicly both governments took pains to declare that the step was undertaken by mutual consent and was entirely in the spirit of their 1971 treaty of friendship. Yet in fact the move came as a shock to the Kremlin. The Egyptian president had for some time been unhappy over the extent of Soviet help for his country and the conditions attached to it. And with the new turn in USSR-US relations he obviously suspected that Moscow would now become even more attentive to American sensitivities concerning the Middle East, and unlikely to assist him in a future military confrontation with Israel.

The Kremlin found itself under the embarrassing necessity of explaining that its new relationship with the US would not lead to its abandoning its old friends and proteges. "Only those politically naive can argue that what we are witnessing is some 'understanding between capitalism and socialism,' its costs to be borne by the Third World," editorialized *Pravda*.[6] On the contrary, detente was bound to help the cause of the developing countries and liberation movements and would establish more favorable circumstances for their struggle against reaction and colonialism. As for itself, the Soviet Union would continue "to rebuff any aggressive attempts by the forces of imperialism" and to render extensive help to the patriots of Angola . . . Mozambique, Zimbabwe, South Africa. . . ." The article went on to quote Brezhnev: "We know very well and we shall not forget that the nations of Asia, Africa, and Latin America form together with those of the socialist camp a mighty detachment of the forces for peace in international relations. Working together, we have already accomplished quite

6. August 30, 1973.

a lot. And we are firmly convinced that in the future too, our paths will not diverge."

A similar reassurance, aimed this time at foreign (and perhaps also some home-grown) Communists, was contained in another *Pravda* editorial entitled "Peaceful Coexistence and the Class War."[7] Here again, for all of the author's skill in dialectics, one can detect a certain ambivalence in the official Soviet position on detente. The article hails recent treaties between the USSR and the German Federal Republic and finds highly encouraging the progress in Soviet-French cooperation. The highest praise, though, is reserved for the Nixon-Brezhnev May 1972 declaration of principles which, it is said, ought henceforth to govern relations between their countries: "The signing of the document marked a veritable breakthrough in the relationship of the two states and set the foundation for the development of their cooperation for the mutual benefit of both sides."

But do all those agreements and rapprochements between the USSR and capitalist powers add up to "a 'retreat' by the Soviet Union and other socialist countries and their acquiescence in the status quo?" Definitely not! "The struggle between the proletariat and the bourgeoisie, between international socialism and imperialism, will continue until the complete and final victory of Communism throughout the world." The policy of peaceful coexistence as pursued by the USSR has been designed to ensure that this "historically unavoidable struggle" takes place "without wars, dangerous confrontations, or an uncontrollable arms race." How then is Communism to conquer the world? "By waging a gigantic struggle for human minds . . . by [Communism] demonstrating its decisive superiority in all spheres of social organization."

It is unlikely that such arguments proved entirely convincing to those for whom they were intended. But for us their importance lies elsewhere. Some Western observers have tried, and on the basis of very meager evidence, to divide Soviet

7. *Ibid.*

leaders into "hawks" and "doves," those who favor militant
policies vis-à-vis the West, and others who believe in moderation and prefer a more cautious method of propagating Communism and enhancing Soviet power. It is more than doubtful
that such division is applicable to the councils of the Kremlin
(and even in the case of American politics it is bound to lead
to gross oversimplifications). The arguments we have just adduced suggest that the basic ambivalence about the advisabil-
ity of "hard" policies versus those more in keeping with the
spirit of detente ran through the mind of the Soviet leadership
as a whole. Detente from the Kremlin's perspective obviously
involved some costs; as already seen in the case of Egypt, some
of the Soviets' clients might decide that since the USSR was
unlikely to promote their interests vis-à-vis the West with the
same vigor, they might as well pursue a more independent
course or even seek their own *rapprochement* with the United
States.

Nor was detente entirely free from an element of risk insofar
as the domestic scene was concerned. But for all such external
and internal inconveniences there were solid benefits to be
garnered from a *rapprochement* with the West. Economic benefits of detente to the USSR became apparent almost immediately following the Nixon-Brezhnev summit. More specifically, the new and friendlier atmosphere in Soviet-American
relations enabled the Soviet government to overcome a serious
agricultural crisis triggered by an unusually bad harvest in
1972, to make up the deficiency in foodstuffs at a relatively
low cost in hard currency, and in the process to have its purchases of large quantities of American grains subsidized by
the US treasury to the tune of several hundred million dollars!

The lesson of the "great grain robbery" of 1972, as it has
become jocularly known, extends beyond economics. The ep-
isode epitomizes the Soviets' ability, displayed on so many
occasions, and not merely when it comes to trade matters, to
compensate by superior negotiating skills for the inherent
weaknesses of their system. And conversely, the story is a not
atypical example of US clumsiness in dealing with the Soviet
Union, regardless of the issue in question, be it a commercial

transaction, as in this case, a political settlement, or an arms limitation agreement.

Agriculture has long been the Achilles heel of the Soviet economy. Before World War I Russia was the world's leading grain exporter (even then an occasional bad harvest could trigger a famine in a given region). But since the Revolution, and especially after the forced collectivization of the thirties, Soviet agriculture has chronically failed to keep pace with the spectacular growth of the industrial sector of the economy, the latter being achieved largely at the price of ruthless exploitation of the peasant. Like other spheres of Soviet life, conditions in the countryside improved considerably following Stalin's death. But though no longer a virtual serf of the state, the peasant, mainly because of the system under which he works, has remained an inefficient producer; even the official statistics estimate his productivity at one-fourth that of the American farmer (in fact the proportion is one to seven or eight). Though greatly expanded, food production has failed to meet the rising needs and expectations of the Russian consumer, needs which post-1953 leaders felt they could no longer afford to ignore. Food shortages in the early 1960s helped precipitate Khrushchev's fall. The events of December 1970 in Poland, when the workers' riots triggered by the increased prices of meat and vodka toppled the leadership of the Communist party, confirmed the Kremlin's concern about political dangers inherent in the food supply question. Unthinkable in Stalin's time, when—as in 1932–33—the government let several million starve rather than expend hard currency on food imports, sizeable grain imports became a regular feature of Soviet trade in the 1960s. The regime preferred to buy from grain exporters other than the US, but the fact remained that only the latter had exportable surpluses large enough to satisfy the Soviets' needs whenever their harvest turned out to be especially low. Occasionally, as in 1964, the USSR did import sizeable amounts of American grain. But otherwise, and for obvious reasons, Moscow was careful in the pre-detente days to avoid even the appearance of being dependent on the US when it came to a commodity as essential as food.

There had been early signs that 1972 was going to be a bad year for Soviet agriculture. Indeed the eventual harvest fell far below the official target, and was 13 million tons smaller than the already not very satisfactory crop of the preceding year (168 versus 181 million tons), when the USSR had been constrained to import considerable amounts of grain, including 3 million tons of livestock feed purchased in the US.[8] The parlous state of Soviet agriculture was evidently absent from the American policy-makers' calculations as they plotted their negotiating strategy for detente. Contrariwise, the fast-approaching food crisis was very likely one of the main reasons for the Kremlin's eagerness to go ahead with the May 1972 summit meeting, even in the face of the recent American escalation of bombing in North Vietnam and the mining of its harbors. Were the meeting to be postponed, the US might well prove unwilling to bail the Soviets out of very serious and pressing economic difficulties.

Within four weeks of the official initiation of detente, a Soviet trade delegation arrived in the US. Negotiating separately with the US Department of Agriculture and several grain dealers, and without revealing the total amount of their intended purchases, the Soviet officials scored a brilliant commercial coup. Once the dimensions of the Russian plight and the general shortage of grain in the world market became known in September 1972, the price of wheat shot up to $2.43 per bushel. By that time, in a series of agreements reached in July and August, the Soviet Union had already contracted for 19 million tons of American grain at $1.61–$1.63 per bushel! Even more amazingly under the Agriculture Department's export price-support program, those sales had been subsidized by the American taxpayer at a cost of more than $300 million.[9] Of course the whole transaction still benefited greatly this country's farmers and its overall balance of trade. At the same time it was largely on account of the Soviet purchase of one-fourth of the US crop that the American consumer found himself paying much higher prices for his foodstuffs.

In viewing retrospectively the Soviets' grain victory, some

8. Marshall Goldman, *Detente and Dollars* (New York, 1975), p. 195.
9. *Ibid.*, p. 212.

have tended to exaggerate its importance. Thus it is clearly excessive to state that "The Soviet Union quite literally had no other choice than to buy our grain or to face mass starvation."[10] Inability to acquire American grain would have led to severe discomfort for the Soviet citizens, perhaps to a few food riots (such as did occur in the early 1960s), a depletion in the country's livestock for some years to come, but hardly to a famine on the order of those which had occurred in Stalin's time. Had the Russians been constrained to pay more, they would have had less hard currency for imports useful for their heavy and defense-oriented industries. But the greatest damage from the American point of view, one must conclude, was psychological. Once again the US had failed to recognize and to bring into play the strength of its assets. That, quite apart from harming this country's bargaining position, was hardly conducive to the mutual respect that must form the basis of sound relations between two great powers. There had been a failure of intelligence, in both senses of the word, on the part of Washington. The experts' warnings about the precarious prospects for the Russian harvest were disregarded by the Department of Agriculture, while the President and his foreign policy advisers were left entirely in the dark insofar as the agricultural picture was concerned.[11] There is added piquancy in the fact that the entire sale was probably illegal in view of a law then on the statute book.[12]

Secrecy and skill in timing its moves have often been the main ingredients of the Kremlin's diplomatic successes. And so it was in this case. The arrival in the US of the first Soviet trade delegation on June 28 was kept secret from the American media and public. The Russians moved quickly to close their deals with several companies, none fully aware of the full scope and urgency of the Soviets' need for grain. The Yankee traders were bested at their own game.

Still one suspects that the US government's sins of commis-

10. Henry Kissinger, *White House Years* (New York, 1979), p. 1270.

11. *Ibid.*, p. 1269.

12. The Agricultural Appropriation Act of 1966 banned the export of government-subsidized food to any country that supplied equipment and war materiel to North Vietnam.

sion and omission concerning the deal resulted not entirely from negligence or quick thinking by the Soviet economic bosses and negotiators. The economic side of detente was important also in Washington's eyes. The idea of the Communists becoming, so to speak, domesticated and easier to live with under the influence of expanded trade and general intercourse with the West had been of long standing. As early as the 1950s an ingenious American sociologist suggested that one way of weaning the Soviets from their unfortunate habits and inculcating in them desirable ones (i.e., similar to those of Western societies) would be to smuggle into the USSR, or even to drop from airplanes, copies of the Sears-Roebuck catalogue! During the 1960s various official American restrictions on trade with the Communist bloc began to be softened and modified. In embarking upon its policy of *rapprochement* with the USSR, the Nixon Administration banked heavily on the beneficial effects of greatly increased Soviet-American trade. The Russians, some influential voices in Washington urged, should indeed be encouraged to seek Western credits and trade, for in view of the increasing addiction of their society to "consumerism," such commercial links were bound to act as a brake on the Kremlin's destabilizing activities abroad. Imperceptibly, it was confidently believed, the Soviet Union would find itself partly dependent on the West for meeting its citizens' expectations of a more abundant life and for Soviet industry's technological progress. Its government would thus be less likely to pursue policies which threatened forfeiting of the West's goodwill and hence a slackening or cut-off of the flow of foreign credits, wheat, and computer technology. Though no longer militarily predominant, the US would still be able, and probably more effectively, to deter Soviet expansionism and aggression through its own and its allies' economic power, or, as the phrase had it, by employing the carrot-and-stick approach toward the Communist bloc.

It would be unrealistic to overlook the fact that such hopes were strongly encouraged by many within the American business community (especially since other Western governments and Japan had for some time now encouraged more trade with the USSR), but ungenerous to explain them solely in terms of *Realpolitik* and profits. It has been a cardinal tenet of the liberal, and hence American, tradition that trade is an important tool in breaking down barriers between nations, in helping to secure peace and international stability.

Considerations such as these may well have played a role in allowing the Russians to score their grain coup. And they undoubtedly were conducive to the Soviet-American trade agreement initialed in Moscow in October of the same year. Under the agreement the US government promised, subject to approval by Congress, to accord the USSR most-favored-nation tariff treatment, thus facilitating the access of Soviet goods to the American market. Both sides affirmed their intention to expand mutual trade during the three years' duration of the treaty to a level much higher than in the past. American firms were to be allowed to share in the construction of the Kama River plant, which was to be upon its completion the world's largest truck factory. (The Soviets' original bid to have it built by an American company had been vetoed some years before by Washington; the main contract went to an Italian firm.) The USSR in turn promised to settle obligations incurred under American Lend-Lease in World War II.

The treaty, as we shall see, was not destined to be ratified. But the years 1972–75 did witness considerable growth in the two countries' commercial transactions. At the height of the Cold War such trade had virtually stopped. (The value of American goods imported by the USSR in 1950 was a derisory $7 million.) In 1971 the Soviet Union imported $162 million worth of American products; the figure grew to $542 million in 1972. The grain purchase contracted the previous summer made the figure shoot up to $1190 million in 1973, and then, as a result of an abundant Soviet harvest of that year, it receded in 1974 to a more typical $612 million.[13] The trend continued throughout the decade.

Both sides' initial expectations of greatly expanded trade because of detente eventually foundered on the political and economic realities of the later years. Still, from the Soviet point

13. Goldman, *op. cit.*, p. 268.

of view, the economic side of detente proved to be of great importance. Though difficulties, mostly political in nature, bedeviled Soviet-American trade after 1974, Soviet commercial dealings with the West as a whole increased considerably. Detente helped loosen the purse strings of the Western and Japanese bankers, and together with the recession of the 1970s whetted businessmen's appetites for Communist markets. Though quantitatively not very impressive, the increase in trade with and availability of credits from the capitalist world provided an important boost to the faltering USSR economy. And indirectly, as has already been suggested, the Soviets garnered substantial benefits, political as well as economic, from the Eastern European Communist countries' expansion of trade with the West. True, when misused, as by Poland, large-scale loans from the capitalists were eventually proved quite counterproductive and contributed to a dangerous economic and political crisis. Still, between 1972 and 1979 the Kremlin could view the overall picture of West-East economic relations with smugness: capitalist credits and technology were succoring the Soviet economy and even more so those of Poland, Hungary, East Germany, etc.

For the West, however, the expected political benefits of expanded commercial intercourse with the Soviets fell quite short of the rosy expectations of 1972. Perhaps if detente as a whole had prospered, trade with the West would indeed have become an important factor in moderating Moscow's foreign policy. But even in the beginning it was clear that the economic power of the US and its allies was not being utilized in a way that would make it effective in altering the basic pattern of the Soviet Union's international behavior, even though paradoxically it did influence (marginally, to be sure) Soviet domestic policies. The carrot-and-stick rationale of East-West economic collaboration was soon exposed as unrealistic, or rather the Soviets went on munching the carrot (and complaining that it was not big enough), while the stick turned out to be but a broken reed. For one thing, the effectiveness of the trade weapon depended, as it still does, on the Western countries and Japan being able to synchronize their economic policies. But even before the onset of the mid-1970s recession, such coordination was increasingly difficult to achieve. And once inflation, growing unemployment, etc., began to afflict the industrial world, it became well-nigh impossible for the Western alliance to present a united economic front vis-à-vis the Communist bloc. If Washington chose to react to some move by Moscow by withdrawing its export license for a highly advanced computer, France or Japan would usually be more than ready to step into the breach and sell Moscow its own version of the American product.

In agriculture the US position has been much stronger than in most lines of industrial goods and techniques; no other country can offer as much grain for export. But the American farmer has balked at being victimized each time his government feels it has to manifest its displeasure over Soviet actions. Western disunity and domestic pressures have thus largely confounded Washington's hopes that increased economic links with the USSR could significantly and positively affect its foreign policy.

It was also too readily assumed that, if carefully monitored, Western trade with the Soviet Union would not serve to enhance its military potential. In an economy as centrally controlled as Moscow's, increased savings and efficiency in the consumer goods sector can without too much difficulty be translated into additional resources for the heavy industry—military one. It would be mistaken to exaggerate the importance of Western trade credits in that or any other connection; total Soviet imports from the "developed" countries in 1976 amounted to $14.5 billion, not a very imposing figure in the context of a gross national product of over $1 trillion. Still economic detente probably contributed to the regime's ability to maintain throughout the 1970s a steady if modest growth in living standards, while at the same time it kept increasing military expenditures at the rate of 4–5 percent per year.[14]

One must not at this point of our narrative try to draw a balance sheet of detente. But perhaps enough has been said to characterize the multiplicity of factors which entered into

14. Goldman, *op. cit.*, p. 275.

Moscow's decision to attempt a new approach to East-West relations, and to suggest that this approach neither represented a cynical ruse nor, on the other hand, sprang from a wholehearted conversion to the cause of international peace and stability.

On the American side, the first fruits as well as disenchantments of detente became apparent with the Vietnam accords finally reached by the US and North Vietnam and initialed in Paris in January 1973. Soviet influence was undoubtedly instrumental in making Hanoi agree to a truce whose provisions fell quite short of what it had hoped to obtain out of the negotiations: not only a US military pull-out from Indochina, but also the replacement of the Thieu regime by a "coalition" government, i.e., a preliminary to a peaceful turning over of the South to the Vietcong. Now the Nationalist government in Saigon would continue, and at this point no one could be quite sure for how long. The US obtained the return of its prisoners of war plus a face-saving arrangement, extremely fragile though it was, in the South. Taking into account the American people's now frantic exasperation with the war and the mounting Congressional opposition to the continuation of American involvement in Indochina, Nixon's administration could thus congratulate itself on achieving "peace with honor." But Washington's gratitude to Moscow for its help in extricating the US from an untenable position was tempered by some bitter reflections: the Soviets, the President and his advisers felt—and probably quite justifiably—could have done much more. No one really believed that the ramshackle arrangement authorized under the Paris agreement would endure indefinitely: enclaves of Vietcong rule and of North Vietnamese troops were to remain within the South, and neither of the Vietnamese contestants could be seriously expected to abide by the truce. When it began its massive intervention in Indochina in 1964 the US government had intended to demonstrate to the world, and especially for the benefit of Communist China, that the "wars of national liberation" did not pay. And indeed the consequences of this one were disastrous to the Vietnamese people, most unsettling for the US, and eventually very disappointing to Beijing. But from the Soviet point of view they proved quite satisfactory.

Though now moderating its tone in talking about the United States, the Soviet press was still allowed to gloat quite a bit over America's discomfiture in Vietnam. "In the name of what has so much blood been spilled and so many billions spent? For what purpose have the invaders inflicted such enormous suffering on a small country ten thousand kilometers away from the shores of America? . . . The results of the war have demonstrated the futility of the imperialist practices of repression and interference in the internal affairs of nations." Another lesson of the war: the "historic victory of the Vietnamese people" was made possible very largely by "the many-sided assistance and support rendered by the Soviet Union."

Russian reporters stressed the human and material losses suffered by the US: "Fifty-three thousand servicemen killed, more than 300,000 wounded, $120 billion expended on Vietnam between 1965 and 1972." *Pravda* and other journals recorded avidly the most plaintive of American avowals of the other, incalculable damage inflicted by the war on the US, that to national morale and self-confidence. A prestigious American journal, noted one Soviet journalist, was forced to acknowledge that US armed might proved unable to subjugate "a tiny [sic] but proud Asian nation devoted to its ideals." Another American paper was said to have editorialized that "the lesson of the war should teach Washington that there are limits to our power and that it can no longer dictate to the undeveloped nations." A noted columnist was cited as confessing that "we are no longer confident that the US must always prevail in international conflicts . . . that money and technology are the decisive factors in war, or that small states will always capitulate rather than defy America's might." "No more Vietnams," proclaimed, allegedly, the *New York Times*, etc.[15]

For all such rather gleeful depictions of America's self-flagellating mood, the Soviet press was still careful not to leave the impression that Vietnam had incapacitated the US as a

15. *Pravda*, February 15, 1973.

world power. The Soviet reader was made aware that the war was hugely unpopular at the home front; the implication was that the story might be quite different in the case of an international conflict in which public opinion stood behind Washington. And the Kremlin was equally careful not to permit any explicit criticism of its partner in detente—the Nixon Administration. On the contrary the President and his advisers were praised for their statesmanship and realism, as exhibited both in their resolve to terminate the unfortunate war and in their readiness to embark on a new type of relationship with Moscow and the entire Soviet bloc. (Various steps in the progressing US-China *rapprochement* were as yet being reported without comment.) The US, as portrayed by the Kremlin to its people, was still a giant, currently a bit wobbly on its feet, but not to be trifled with.

Detente, then, in the view of the Soviet leadership, was premised on a considerable respect for America's power and for its ability, once it extricated itself from the Vietnamese morass, to act decisively in the international arena. Hence the USSR would observe a considerable restraint in its policies. Not to the point, of course, of meekly acquiescing in the status quo. The Soviets, Khrushchev had told John Kennedy in Vienna in 1961, could not be expected to sit with their arms folded and just watch what was happening in the international arena, a sentiment undoubtedly shared by Brezhnev and his colleagues. But they would be careful not to endanger detente by policies that might threaten to provoke a vigorous American reaction.

"The recently achieved and positive changes in the international situation must be made irreversible," proclaimed a resolution of the April 1973 session of the Central Committee.[16] And a lead editorial in the official organ of the Party developed still further the rationale of the new, more prudent line of the Kremlin's foreign policy. The editorial included, to be sure, the usual boast that the configuration of forces in the world arena was changed greatly in favor of the Communist camp, and repeated the ritualistic phrase about the imperialist

16. *The Communist*, no. 7 (May 1973), p. 1.

powers now realizing that a nuclear war would be suicidal for capitalism.[17] But then there followed an explicit acknowledgment that such a war would be catastrophic for all. "History has fully demonstrated that the policy of peaceful coexistence is not a tactical manoeuvre, but a basic necessity, for in our days it has become the only alternative to a *thermonuclear catastrophe.*"[18]

Again, a rare (in a Soviet official organ) admission that a nuclear war would be disastrous for both sides. Such an avowal has usually been intended to signal to the initiated that the "objective conditions" dictated a policy of restraint and caution—or, to use the vernacular, the Kremlin was ready to pull in its horns. What was the concrete nature of those intended offerings on the altar of coexistence? Was the USSR going to use its influence in the Arab world to try to lower the level of tension in the Middle East? Would it abandon its efforts to establish footholds in various fragments of the crumbling Portuguese empire in Africa? Would Moscow continue to preach moderation to Hanoi, and keep a rein on Castro's ambition to become the Lenin of Latin America? We can only speculate about answers to such questions. For whatever the Soviets' original image of detente, there soon ensued a series of events that altered the Kremlin's perception of the cautions and constraints it would have to observe because of its new relationship with the United States.

Detente was still in its infancy when an unexpected calamity befell one of its parents. With the Watergate affair erupting in the spring of 1973, the attention of the American public and the energy of its government became increasingly absorbed by the political drama at home. The American system offers no alternative to presidential leadership in foreign policy. A presidency under siege, with its incumbent struggling for political survival, must of necessity have far-reaching and deleterious effects on US standing and freedom of action in the world at large.

Prior to Watergate the Soviet leaders' private assessment of

17. *Ibid.*, p. 7.
18. *Ibid.*, p. 9. My italics.

Nixon may well have diverged somewhat from that conveyed to them by Kissinger (at the President's explicit instruction!) which characterized him as "direct, honest, strong, fatalistic . . . not affected one iota by public opinion." Still everything indicates that Brezhnev and his colleagues saw Nixon as an effective policy-maker, i.e., one who could deliver on his promises, and, by the same token, a man who would react vigorously were the Soviets to break theirs. And so any drastic change in the President's political fortunes had to have as a corollary the Kremlin's reassessment of the benefits and costs of detente.

To be sure this reassessment was slow in coming. Moscow's first reaction to the political melodrama now unfolding in Washington must have been one of bewilderment: even though the Politburo had a considerable number of experts who offered quite sophisticated advice about the strange customs of American politics, the men in the Kremlin must have found it incredible that a thing as trivial, and in a way so natural, as a little spying on one's opponents should throw the whole machinery of the US government into turmoil and threaten to topple a man just elected by a landslide vote. Surely there must have been other and more substantial reasons behind what, to the Soviet mind, suggested a conspiracy against the President; perhaps the very fact that it was he who had initiated the *rapprochement* with the USSR. As yet the Kremlin was confident that the whole hullabaloo would not amount to much. The deputy attorney general of the Soviet Union, Malyarov, expressed the prevalent official view when he remarked that the Watergate business was "demagoguery . . . all for show . . . [and] would not amount to anything provided Nixon displays firmness."[19] Being deeply conservative by temperament and set in their ways, the Soviet leaders have usually preferred the foreign affairs of other powers to be managed by those whom they had come to know, if not ex-

---

19. *A Chronicle of Human Rights in the USSR* (New York), June–August 1973, p. 18. Malyarov made these remarks in his interrogation of the noted physicist and dissident Andrei Sakharov.

actly trust (a rare sentiment on the Soviet Olympus), and whose reactions to their own policies, they believed, could be gauged with some accuracy.

The mounting domestic crisis in the US was thus not allowed to delay the scheduled state visit by Brezhnev. The Soviet statesman arrived on these shores on June 17, and the new relationship between the two countries was underlined by the special treatment accorded to the General Secretary and his entourage. Unlike Khrushchev on his tour of America in 1959, Brezhnev was not subjected to a hectic round of sightseeing, public appearances, and sometimes acrimonious press interviews. To be sure this reflected to some extent the difference in the two men's personalities: Brezhnev reserved and dignified, his predecessor boisterous and populistic. But there was also a studied effort to impart to the occasion a solemn character, to present it as a momentous encounter between two leaders whose intimacy was of supreme importance to the relations of their countries and to world peace; a scenario that on the President's side may have been influenced by the desire to distract the public's attention from his domestic troubles. The Soviet guests were accorded the unusual courtesy of a weekend by themselves at Camp David, thus allowing them, unlike the Americans one year before in Moscow, a period of reflection and of recuperation from jet lag before plunging into negotiations. Following the latter, and again as if to emphasize the personal element of the visit, the two potentates flew together to the President's residence in California—increasingly Nixon's refuge from what to him was the pestilential atmosphere of Washington, with its Congressional hearings and never-ending stream of revelations about Watergate and other indelicate features of his administration.

Brezhnev's televised broadcast to the American people stressed the importance of personal links between the leaders of the two countries. He was highly complimentary about Nixon, whose "administration united his efforts with ours to lead Soviet-American relations to a really new path."[20] He and

20. *Pravda*, June 25, 1973.

the President had decided, Brezhnev announced, to continue their meetings on a regular basis, the Soviets "looking forward to President Nixon's visit to the Soviet Union next year."

Aside from thus trying to help a fellow politician in trouble, the General Secretary took pains to reassure the American public that detente was not a one-way street, but of great potential benefit to *both* nations. The Nixon-Brezhnev declaration on the basic principles of relations between the US and the USSR, signed one year before in Moscow, was not mere rhetoric, but something very concrete that the Soviets "have already started to implement, and firmly intend to go on implementing." Expanded trade between the two countries was another demonstration of the mutual benefits derived from detente.

Apart from frequent and laudatory references to Nixon in his speech, Brezhnev gave other indications that the Russians expected the current administration to survive its domestic troubles and to be able to act vigorously on America's behalf in international affairs. A joint communiqué expressed confidence that further and permanent agreement to limit strategic offensive arms could be reached by the end of 1974, presumably on the occasion of the President's visit to Moscow. Both sides referred to the situation in Indochina in terms implying that the Russians believed the attempt to establish complete Communist control of the area would have to be deferred for a considerable time, if not indefinitely.

All in all, as of June 1973 the Moscow leadership was still far from realizing to what extent America's internal crisis would cripple its ability to act effectively in world affairs and by the same token offer new opportunities and almost irresistible temptations to the Soviet Union.

However, as the year progressed, the Kremlin could not fail to note that Nixon's political base was eroding. What had begun as a Washington scandal was taking on the appearance of a major constitutional and political crisis. The country had just begun to recover from the malaise induced by Vietnam, and a new national consensus on foreign policy had just begun to emerge when Watergate threw the relations between the executive and Congress into fresh disarray, making it virtually impossible for the US to act in a concerted and convincing manner in meeting a foreign emergency. And on another level, preoccupied as it was by the domestic drama, the American public could not be expected to pay much attention to the portents of danger abroad: the implications of the breakdown of the Portuguese colonial empire in Africa, the fact that the Paris agreements had failed to stop fighting between the Saigon regime and the Vietcong, and the rising tension in the Middle East.

It was in the last-named area that detente was put to its first major test. And it is no exaggeration to say that the Arab-Israeli ("Yom Kippur") war and its corollaries demonstrated the fragility of the premises on which the recent *rapprochement* between the two superpowers had been based. Or, to put it differently, what happened during, and especially as a consequence of, the war helped convince the Kremlin that detente did not put it under any special constraints insofar as Soviet foreign policy was concerned. The events of October 1973 vividly demonstrated that not only American politics but the Western alliance was currently in disarray. Thereafter there could be but little reason for the Soviet Union to refrain from undermining Western influence and interests in the Third World, while insisting that such activities should not be allowed to interfere with friendly and cooperative East-West relations.

The Soviets had long known about President Sadat's determination to avenge Egypt's defeat and humiliation in the 1967 war. Their attitude toward the Egyptian leader had to be ambivalent. He was not, as shown by his actions in 1972, a friend of the Soviet Union. Yet to deny his pleas for arms could have meant Egypt's moving into the pro-American camp, and thus perhaps a decisive setback to the Soviet position in the Middle East, built at so much effort and expense ever since 1955. (Sadat cleverly teased Moscow by open diplomatic contacts with the Nixon Administration and also by secret—though probably known to the Soviets—negotiations through Kissinger that began about the same time he expelled most of the

Soviet military advisers from Egypt.) And so, perhaps with some private sighs, the Russians resumed in January massive shipments of arms to Cairo.

In our present state of knowledge we can only conjecture about the actual date on which the Kremlin learned of the impending Egyptian-Syrian attack on Israel. A direct notification that an attack was imminent, but without specifying the actual day, was conveyed to the Russians by the Egyptian authorities, according to the latter on either October 1 or 3.[21] On the fourth and fifth the Soviets proceeded quite openly to evacuate their diplomats' families and civilian advisers from Egypt, something which, again, they would not have done had they been really intent on not betraying the secret. Even so the American and Israeli services must have been asleep, for when the Egyptians and Syrians struck on the sixth, they achieved complete surprise. A passage in the Nixon-Brezhnev declaration of 1972 (the very declaration that the General Secretary had recently proclaimed to be "not a mere statement of good intentions, but a program of vigorous and consistent actions") placed both governments under an obligation "to do everything in their power so that conflicts and situations do not give rise to increased international tensions." One interpretation, i.e., American, of the pledge held subsequently that it obligated the USSR to notify and take counsel with its partner in detente the minute it knew that an attack on Israel was impending. To justify themselves the Soviets could have pleaded that the situation in the area had been explosive for a long time, and was it up to them to teach the CIA how to go about its job? To betray Sadat's confidence to Washington would in all likelihood have meant tipping off the Israelis, and thus depriving the Egyptian attack of the main ingredient of its success—surprise. It is quite likely that the Kremlin believed it discharged its detente duties by oblique warnings. Speaking before the UN on September 25, Foreign Minister Gromyko (recently elevated to the Politburo, hence now a policy-maker

21. Mohammed Heikal, *The Road to Ramadan* (New York, 1975), p. 24. The third is the day given by Sadat himself, the first by his then close adviser, Heikal.

as well as the official spokesman) said about the Middle East, "The fires of war could break out onto the surface at any time and who can tell their consequences?"[22]

It is thus hard to sustain the version of these events that holds that the Soviets actively encouraged Sadat in his venture. Rather the evidence points to their playing a double game, on the one hand solicitous to preserve their influence in the Arab world, on the other fearful of damaging their fresh ties with the US. Such at any rate must have been the conclusion of Sadat himself; despite massive Soviet material help and diplomatic support, his subsequent actions demonstrated that he did not consider the USSR a loyal ally. And perhaps with equal justification the Kremlin would see the forceful Egyptian leader as one who used the Soviets for his own purposes rather than a reliable instrument of their policies.

It is clear, however, that the Soviet attitude underwent a considerable change in the beginning of the conflict and became again one of strong support for the Arabs. The very first Soviet reaction reflected both surprise at the initial successes of the Egyptians and Syrians and considerable skepticism as to their ability to prevail once Israel recovered from the unexpected shock and mounted a counteroffensive. Inevitably the war was presented to the Soviet public as having been started by Israel. "Israel has attacked Egypt and Syria," was the headline in *Pravda* on October 7. The next day's official statement was still careful not to commit the USSR to any concrete course of action. The Soviet Union was declared to be "a dependable friend of the Arab states." As such, "it has condemned the expansionist policies of Israel and supports firmly the legitimate demands for the liberation of Arab territories seized by Israel in 1967."[23] Significantly the statement refrained from any optimistic assessment of the Arabs' chances of repelling the "aggression."

Late on the sixth the Soviet ambassador sought out President Sadat and suggested, to the latter's amazement, that now that Egypt and Syria had destroyed the myth of Israeli invin-

22. Galia Golan, *Yom Kippur and After* (New York, 1977), p. 68.
23. *Pravda*, October 8, 1973.

cibility the Soviet Union should promote the idea of an immediate cease-fire. The next day the Soviet envoy informed
Sadat, evidently mendaciously, that Syria's president had told
Brezhnev that his army was beginning to experience difficulties and that he would be in favor of the USSR placing a cease-
fire resolution before the Security Council.[24] Israel had been
humbled; Egypt and Syria should stop while they were ahead,
the ambassador argued. Had the Egyptians agreed, the Soviet
Union thus would have demonstrated its fidelity to the Arab
cause, while avoiding a clash with the US. These evidently
were Soviet motives in proposing an end to the fighting. But
by now Sadat was not content just to cross the Suez Canal and
push the enemy a few miles back. He craved a major military
and then a diplomatic victory and did not want a premature
cease-fire to rob him of them.

The Kremlin thus confronted a difficult dilemma. It had to
be apprehensive of a decisive victory by either side. Another
Arab defeat would in all likelihood spell an end to Soviet influence in the Middle East and have unpleasant reverberations
elsewhere. A threat to Israel's existence posed by Egypt's
triumph would almost inevitably bring an intervention by the
US and thus threaten a confrontation between the two superpowers. To ward off the first and, in their view, likelier contingency, the Soviets on October 9 began a massive airlift of
arms to Egypt and Syria, whose army was already beginning
to falter under the weight of Israeli counterattacks. And in a
message to President Boumedienne of Algeria, intended to remain secret, Brezhnev called upon the other Arab states to
come to the aid of their embattled brethren. The burden of
military help to Egypt and Syria and of warding off too drastic
a setback to their armies was thus to be shouldered by their
Moslem allies, while the USSR confined itself to the role of a
diplomatic cheerleader for the Arab cause. Simultaneously the
Soviet leader exchanged messages with Nixon stressing their
mutual determination to preserve detente; this communication took place openly and through the normal diplomatic
channels rather than through the "hot line," to emphasize that

24. Heikal, op. cit., pp. 209–12.

what was happening had not created a serious crisis in Soviet-American relations.

By October 12, however, the Kremlin had jettisoned its hitherto prudent attitude and was ready to commit itself more strongly to its embattled allies. There were several cogent reasons for the Soviets' reassessment of the situation and their conviction that this new stand would not bring a dangerous confrontation with the US.

Moscow was undoubtedly encouraged by Washington's hitherto indecisive reaction to the crisis and could not fail to attribute it, at least partially, to the unprecedented chaos of the American political scene. With Vice President Agnew having just resigned and the now quite open talk about impeaching the President himself, the Soviet leadership could be excused for believing that the administration was in no position to act forcibly abroad or to abandon detente. America's allies had already indicated that they would not support Washington in any steps that might bring upon them the wrath of the Arab producers of oil. The Western European states had long been critical of what they saw as America's excessive identification with Israel's cause, their apprehension being based on their dependence—in some cases almost total—on oil from the Middle East.

And indeed very early in the conflict the Arab members of OPEC signaled their intention of employing the oil weapon to strike at those who helped Israel. Several of them soon cut off petroleum shipments to the US and the Netherlands, and by October 17 they all agreed to reduce their production 5 percent each month until the Arabs' principal goal—Israel's evacuation of the territories it occupied in 1967—was accomplished. The genie was out of the bottle: the oil cartel, OPEC, hitherto not very effective, now cast a growing, and, to the West, ominous shadow over the international economic and political scene. Its Moslem members became able to exercise world-wide influence grotesquely out of proportion to their military strength and internal stability.

The Kremlin was thus presented with opportunities and temptations it could not resist. A strong pro-Arab stand on its part carried little risk while promising substantial dividends, and not merely in the Middle East. The war exposed the Achilles heel of the Western alliance and Western economy: oil and its political ramifications. By helping Egypt and Syria preserve at least a semblance of military victory, the USSR could do more than undo the consequences of its lamentable performance in the 1967 crisis: even conservative Moslem regimes like those of Iran and Saudi Arabia that could not be expected to become pro-Soviet would now adopt a much more independent and demanding attitude towards the United States. The war, which at first posed so many knotty questions for the Soviet leaders, after a week or so must have appeared to them as a veritable bonanza.

One thing could still spoil the new and pleasing prospects: a reversal in the fortunes of the war and a defeat for Egypt and Syria. To guard against this possibility the USSR greatly expanded the airlift of war materiel to its clients. The Soviet fleet in the Mediterranean was being strengthened. And it was on October 12 that some Soviet airborne divisions were put on alert.[25]

If some in Moscow's ruling circles were apprehensive that all such moves were bound to bring countermeasures by the US, they might well have been reassured by an item featured prominently in *Pravda* on October 15. It presented the energy situation in the US as critical. The President himself had warned that serious shortages must be expected during the coming winter and that in all likelihood US industrial production would have to be curtailed and many schools and hospitals deprived of heat might have to be shut down. It was estimated (*Pravda* did not say by whom) that the present US reserves of unrefined oil could last but twenty days, those of gasoline forty. Perhaps it was a coincidence that this somber (but for Moscow, in view of its current moves, comforting) account was printed next to the extensive report on the Middle East war.

But as in the previous Israeli-Arab confrontations, the whole

25. Karen Davisha, *Soviet Foreign Policy Towards Egypt* (London, 1979), p. 68.

politico-military setting of this war suddenly and sharply changed. Whatever the external or internal reasons for its initial indecisive reactions, the US government by October 14 authorized and began a massive arms airlift to Israel. And on the same day the Egyptian armor driving toward the Sinai mountain passes was thrown back. With the bulk of their foe's forces committed and stalled in the peninsula, the Israelis now moved to cross the canal. By the sixteenth they had a sizeable bridgehead on the left bank, thus threatening to encircle an Egyptian army.

These developments must have alarmed the Soviets and re-awakened the fear that another 1967 might be in the offing. Prime Minister Kosygin rushed to Cairo to try to persuade Sadat once more that a cease-fire was in order. Back in Moscow on the nineteenth, Kosygin must have convinced his Politburo colleagues that speedy action was now essential lest (a) Egypt's and Syria's military positions suffer irreparable damage; (b) the Soviet Union be placed between the alternatives of acquiescing in the Arab defeat or risking a dangerous clash with the US. Moscow sought urgently a conference between the two superpowers' foreign ministers. On the twentieth Kissinger, by now secretary of state as well as security adviser to the President, was conferring in the Kremlin. Not even the Cuban crisis of 1962 had produced so much hectic diplomatic activity between the two powers. Both now agreed on the necessity of a cease-fire, with the belligerents keeping their military positions as of the moment. The US and the Soviet Union jointly sponsored Security Council resolution to that effect was passed unanimously, China abstaining, on October 22. The resolution called for the cease-fire to be followed by negotiations aimed at establishing a durable peace and implementing the 1967 UN vote that had called for Israel's withdrawal from the territories it conquered in the Six-Day War.

On his way back from Moscow the secretary of state stopped in Tel Aviv. Here, according to the subsequent Soviet allegations, Kissinger ran into bitter Israeli objections to being asked to stop fighting precisely when the fortunes of war were changing in their favor. The secretary, continued the Soviet version, insinuated to his hosts that while formally accepting the cease-fire they did not have to stop fighting right away.[26]

Fighting indeed did continue; by the twenty-fourth the Israelis had completed the encirclement of the Egyptian Third Army, posing a threat to Cairo.[27] Brezhnev found himself a recipient of urgent messages from Sadat. And the Soviet Union was already being widely criticized in other Moslem states for sacrificing the Arab cause on the altar of detente and for not insisting that the cease-fire be accompanied by a definite understanding that Israel withdraw to its pre-1967 borders. There took place now a new and risk-laden shift in the Kremlin's stance. After being rebuffed in his plea that American and Soviet military contingents police the truce jointly, Brezhnev on the twenty-fourth notified Nixon that the USSR would be compelled to act unilaterally to stop the Israelis from violating the cease-fire. At the same time Washington learned that several Soviet paratroop divisions had been placed in combat readiness. On the twenty-fifth the US put its armed forces on world-wide alert.

There are good reasons to believe that both sides were bluffing, but there was still an undeniable danger that the situation might get out of hand. Paradoxically, both bluffs succeeded: the Soviet threat led the US to step up its pressure on Tel Aviv to desist from further fighting. And instead of thousands of parachutists, only seventy Soviet military observers descended on the battlefield to monitor the cease-fire. As of the twenty-sixth the Middle Eastern crisis reverted from this violent phase, which had begun on October 6, to its "normal" festering condition.

Both in Moscow and Washington the leaders congratulated themselves on overcoming the most dangerous crisis since 1962 and credited this achievement to detente. Yet it is doubtful that the scenario would have been much different had the crisis erupted before 1972. Indeed some in the US questioned Nixon's motives in ordering the alert, uncharitably attributing

26. Y. M. Primakov, *The Anatomy of the Near Eastern Conflict* (Moscow, 1978), p. 283.

27. Walter Laqueur, *Confrontation* (London, 1974), p. 173.

it to his domestic predicament, rather than the foreign emergency. The President was in the middle of a battle over those famous White House tapes, and was currently encircled by Congressional and journalistic critics perhaps more thoroughly than was the Egyptian Third Army by the Israelis. Hence, insinuated the unkind observers, he needed a dramatic gesture to effect a breakthrough insofar as American public opinion was concerned. The alert did not call for the highest order of war readiness, and what the US would have done had the Soviets come through on their threats and sent their troops to the Middle East must remain a matter of conjecture.

It is easier to speculate on what the Kremlin had in mind: primarily what its warning did in fact achieve. Had the US pressure on Israel been unavailing, Soviet troops would in all likelihood have come to the Middle East to establish a protective shield around Cairo, but not to fight unless attacked.

The war underlined the peculiar danger to world peace posed by the Arab-Israeli problem, a danger as serious now as it was in 1973. In this area the two superpowers' commitments to the opposing parties are so complex and inextricable that neither can predict or fully control its course of action once the festering conflict erupts into actual fighting. But in the immediate wake of the war both parties emerged with a certain bitterness towards their superpower allies: the Israelis blamed American pressure for robbing them of a victory which seemed within their grasp; the Arabs felt resentful that the mighty USSR did not, except verbally, engage itself wholeheartedly on their side, and that Egypt's and Syria's initial successes in destroying the myth of Israel's invincibility were not crowned by the latter being compelled to give up its conquests of 1967.

Detente survived the October war, but from now on it would be in the Kremlin's view a different kind of detente, one requiring fewer constraints on its part. After a period of indecisiveness and confusion, the US had finally displayed some resolution in coping with the Middle East emergency. But would its hobbled governmental machinery be capable of reacting speedily and effectively, would its public opinion sanction similar action in a crisis in an area where Americans' emotions and interests were less intensely involved, say in Africa? What would have been inconceivable a few years before did happen in 1973: the NATO powers dared to say no to Washington and refused to let their facilities be used for the American airlift of supplies to Israel.

"During the October war the Arab states used for the first time the mighty weapon of oil," notes a Soviet author contentedly. "They reduced their production by 30 percent from that of September 1973. At the same time a complete embargo on oil deliveries was applied to . . . the US and the Netherlands and as a consequence the Rotterdam refining and distribution center supplying the entire EEC was paralyzed. Another result of the 'oil war' carried out by the Arab world was the drastic rise in oil prices. . . . By the end of the year they were quadrupled. Such measures struck forcibly at the interests of all capitalist countries, primarily at those of Western Europe and at Japan, 80 percent of whose oil imports come from the Arab states . . . and even in the US the Arab 'oil sanction' served to deepen and complicate the energy crisis."[28]

The Kremlin had good reason to be pleased by the broader reverberations of the Middle East crisis. It weakened the Western alliance and deepened the division within it. The cautions and constraints of detente, insofar as the Soviet side was concerned, had been dictated largely by the economic strength of the United States and its allies. The flourishing state of the capitalist economies, their prodigious and uninterrupted growth during more than two decades, had hitherto stood as a vivid refutation of everything Marxism and Leninism preached. The growing prosperity and industrial strength of the West also had concrete and—from Moscow's point of view—undesirable effects. It weakened the appeal of Communism in the Third World and threatened the rationale of Russian rule in Eastern Europe.

The Arabs' successful use of the oil weapon helped to break the charm and struck at the vulnerable spot in the West's

28. Primakov, op. cit., p. 288.

economy. Now OPEC as a whole had learned the strength of its monopoly position and would exploit it uninhibitedly. Rapidly rising energy costs and the vast transfer of capital to the oil-cartel countries played increasing havoc with the entire economic and social mechanism of the non-Communist world. Inflation, henceforth its constant feature, in turn triggered intermittent recessions, stunting the economic growth of the US and Western Europe. We already noted how even during the fat years the economic strength of the West had never been adequately translated into political power. Now recession and its political consequences made it even more difficult for America and its allies to act in unison and to develop effective policies either vis-à-vis the USSR or on other issues.

The West's apparent helplessness in the face of OPEC's exactions was all the more startling if one considered the fact that most of its member states were dependent on the US and its allies insofar as their external security and the internal stability of their antiquated regimes were concerned. (Only the Iranian Revolution of the late seventies damaged the latter assumption.) In view of their fear of Communism, it was only from the West that Saudi Arabia and Iran, not to mention such mini-states as Abu Dhabi or Qatar, could obtain military and industrial supplies; it was only there that they could invest their recent profits, find experts, and train their own people in the skills required by modernization of their societies. Had the industrial powers been able to coordinate their economic response to the OPEC challenge (never mind any military measures and threats), the oil cartel would have been broken or at least compelled to moderate its exactions. But even here, on an issue which concerned so directly the well-being of their citizens, in a situation which posed no danger of war, the Western nations proved incapable of any concerted action that would have prevented this motley assembly of shaky authoritarian regimes, feudal monarchies, and petty sheikdoms from obtaining a debilitating grip on their economies. The Nixon-Brezhnev declaration of May 1972 had pledged both sides not to seek "unilateral advantages at the expense of the other directly or indirectly," but certainly detente did not obligate the USSR to refrain from benefiting from the West's self-inflicted wounds!

The oil imbroglio and other discomfitures suffered by the US and its allies were presented by the Soviet press as resulting from the ineluctable forces of history, as well as being a just retribution visited by the Third World nations upon their former imperialist masters and exploiters. To be sure the Soviet government was not reluctant to assist those forces of history: it applauded the use of the oil weapon by the Arab states and urged them to nationalize oil fields and production facilities still owned by Western companies. Nor did the Soviet Union disdain to cash in on the commercial opportunities presented by the crisis: it increased its sales of oil and gas to the European states struck by the Arab embargo, something that was noted and commented upon caustically in a number of Middle Eastern capitals.

Disillusionment with the Soviets' behavior in one of them, at least, was to have long-lasting effects on the politics of the whole region. Indebted though he was to the Soviets for saving Egypt from a possible military disaster, President Sadat had now established a new, more balanced approach in foreign policy. In November Cairo announced that it would resume diplomatic relations with the United States. It was mainly through American mediation that the Israeli-Egyptian ceasefire was confirmed on November 11, and then on January 18, 1974, a definitive disengagement of the armed forces of the two countries was brought about. American diplomacy, more specifically Secretary Kissinger's indefatigable labors, was credited by Cairo (and to some extent blamed in Jerusalem) with enabling Egypt to claim the honors of the battlefield: under the disengagement agreement the Israelis evacuated their bridgehead on the western bank of the Canal and pulled back in Sinai; the demilitarized zone between the two armies was to be patrolled by a UN force.

Fighting had subsided, but real peace in the Middle East remained as remote as ever. One would have thought that in the wake of the war the Kremlin would have felt it advisable to reduce its commitments in the region. The overall world situation now opened up tantalizing opportunities for the Soviets in other parts of the Third World. Why then maintain a high degree of involvement in the one area to which the sensitivity of American public opinion was so great, and where consequently no amount of domestic economic and political trouble would prevent the US government from acting forcibly, even at the risk of a confrontation with the USSR? Why not allow the US to try to pursue the virtually impossible task of trying to reconcile the Arabs and Israel, incurring in the process resentment on both sides, as well as fresh troubles with European allies?

Yet though there must have been voices urging such a course of action, the idea of even a partial withdrawal from an area where its influence and prestige have once been massively projected has always run against the grain of the Soviet Union's foreign policy. Were the Soviets as Machiavellian as they have often been said to be, they might well have concluded that they had already reaped the maximum possible benefit from the Middle Eastern question by helping to turn a region once firmly within the Western sphere of interest into a veritable thorn in the side of the capitalist world. But skillful and pragmatic as Soviet foreign policy has been, its makers have never been able to emancipate themselves completely from some (to an outsider often irrational) dogmas. And one of these has been the conviction that the very term "withdrawal" did not belong in the Soviet political lexicon. Withdrawal from any major international arena was a symptom of a state's decline, and hence eventually of the decadence of its political system as well. Britain and France had given up their empires, and not even de Gaulle's eloquence covered up the fact of their thereby greatly reduced standing in the world. The United States' prestige was obviously and seriously damaged by its withdrawal from Southeast Asia. Without the debacle of Vietnam it would have been inconceivable that a US client such as the Shah of Iran could have become one of the principal instigators of OPEC's price rise.

American diplomatic successes in the Middle East, where

Secretary Kissinger's mediation procured first the Israeli-Egyptian truce and then, five months later, a similar disengagement agreement between Israel and Syria, were thus bound to arouse deep apprehension in Moscow. The real aims of US policies in the area, wrote a Soviet expert, were not so much to achieve peace as "(1) to isolate the Arabs from the USSR; (2) to improve Israel's military position, damaged by the October war; (3) to divide the Arab states, thus undermining their strivings towards unity against imperialism, especially those efforts touching on the oil question." [29]

It is then clear why the USSR redoubled its own efforts not to be pushed out of the picture. As a bow to detente, the Soviets agreed to co-chair with the US a peace conference on the Middle East, which opened in Geneva on December 21, 1973. They did so despite, or perhaps because of, the realization that the conference was bound to fail. And indeed in view of Israel's reluctance to discuss the issue of frontiers and of Egypt's insistence, seconded by the Soviet Union, that Israel must return *all* the territories it had seized in 1967, Geneva was to become on this issue, as on so many others before, a synonym for diplomatic futility.

Throughout the year following the October war Moscow acted vigorously to recoup its standing within the Arab world. Over and above its efforts to mend its relations with Egypt, now quite clearly manoeuvering between the two superpowers, the USSR acted to strengthen its ties with the more militant members of the Arab world, notably Syria and Iraq. A breakthrough of some consequence was achieved by the Soviet *rapprochement* with Libya. Its volatile young dictator, Colonel Khaddafi, had hitherto been equally anti-Soviet and anti-Western in his policies,[30] his attitude reflecting the traditional orthodox Moslem hostility toward Communism. But the Libyan leader had become aggrieved by Egypt's consent to a truce with Israel. The fact that the former now gravitated to-

29. Primakov, *op. cit.*, p. 299.

30. Libya purchased some Soviet arms in 1970, but after the attempted Communist coup in Sudan in 1971, Khaddafi became very critical of the Russians.

ward the US prevailed over Khaddafi's religious scruples. In May 1974 he sent his prime minister on a mission to Moscow. A joint Soviet-Libyan declaration condemned "Israel, Zionism and the imperialist forces supporting them" and pledged cooperation between the two countries in several fields.[31] Libya, with its population of two million, became, because of its oil wealth, a factor of considerable importance on the international scene. Its vast hard-currency earnings enabled it to purchase the most up-to-date arms from France and, beginning in 1974, the USSR; its eccentric leader also assumed the role of paymaster for a variety of world-wide terrorist organizations. In the topsy-turvy Middle Eastern politics of the seventies, one could no longer draw a firm line between the more conservative (hence presumably anti-Communist) Moslem states and the radical ones susceptible to Soviet blandishments.

In its search for new friends, Moscow was now ready for closer and open association with the Palestine Liberation Organization. On his previous visits to the USSR its leader, Yasir Arafat, had been hosted by the "unofficial" Afro-Asian Solidarity Committee, and his contacts with the Soviet leadership were not publicized. In 1974 Arafat's reception in the USSR (he went there twice) bore a more formal character, and the Soviet government indicated its readiness to consider the PLO as the legitimate representative of the Palestinian people.

The Palestinian issue now became increasingly the main prop of Soviet policy in the Middle East, and cooperation with the PLO one of its chief instruments. Hitherto the Kremlin had judged a closer association with the militant Arab organization as inappropriate in view of its terrorist methods and the reverberations they produced in the West. The PLO's refusal to acknowledge Israel's legitimacy was also a barrier to its official links with the Soviets, who for all the violence of their attacks upon Zionism (some not free from anti-Semitic undertones) had never questioned the right of the Jewish state to exist.

The new closer linkage with the PLO did not, according to official interpretations, represent a repudiation of the Krem-

31. *Pravda*, May 20, 1974.

lin's previous position. Rather, it was asserted, the Soviet
Union was trying to instill respectability and moderation in
its partner, and convince the PLO, on both moral and practical
grounds, to desist from methods and postulates that could not
benefit its cause: "Soviet spokesmen in their contacts with the
Palestinians explained to them the futility and extreme harm
which individual terror does insofar as the task of national
liberation is concerned. . . . Terrorism is as irreconcilable with
moral principles as it is counterproductive from the theoreti-
cal and political points of view."[32] In any case Moscow's official line held that acts of terror against Israeli civilian popu-
lations had been perpetrated by extremist splinter groups, and
in contravention of the Palestinian organization's policy.

Whatever the sincerity of such protestations, there is no
question that the USSR felt it to be in its own, as well as the
PLO's, interest for the latter to assume a more respectable and
moderate public image. Soviet espousal of the Palestinians'
right to their own homeland and state was bound to find wide
acceptance, and not merely within the Arab world; it would
prove a potent challenge to America's announced policy of
seeking peace "step by step," rather than by addressing the
territorial issue as a matter of principle right away. The effect
of the Soviet gambit would have been much strengthened if
the PLO had abandoned its uncompromising stance concerning Israel. There could have been, and in fact would be, considerable support for a Palestinian state within the territories
wrested from the Arabs in 1967—i.e., the West Bank of the
Jordan and the Gaza Strip (with Sinai returning to Egypt)—
but not for a solution implying that Israel would be erased
from the map. But in urging the PLO to observe, or at least
profess, moderation, the Kremlin ran into a dilemma not un-
common when dealing with a militant nationalist movement:
its leaders' fear lest even by feigning moderation they be denounced and displaced by the more extreme among their followers. The enemies of any compromise with Israel have always felt that they had a prescriptive right to Soviet help; if
they chose to be moderate they would not need the Russians.

32. Primakov, *op. cit.*, p. 133.

The moderate Arabs have customarily put the US on the spot by asking, in effect, "What are you ready to do for us?" and the same question has been addressed by the hard-liners to Moscow. There have been other embarrassing queries asked of the Soviet leaders: why were they, professed champions of the Arab cause, allowing thousands of Soviet Jews to emigrate, most of them settling in Israel and strengthening the ranks of Zionists? Like almost everyone studying the Middle Eastern scene, one Soviet commentator was drawn to some melancholy reflections: "The entirely negative attitude of the Israeli leadership toward the [Palestinian] question has barred the path to a solution of the conflict and has served to perpetuate extremism within the Palestine resistance movement."[33]

The convolutions of Middle Eastern politics cast a shadow over detente just as they influenced other dimensions of Soviet foreign policy. As to the former, both Moscow and Washington kept reassuring each other and the world that their differing approaches to the Arab-Israeli conflict had not been allowed to endanger their *rapprochement*, and that their collaboration, though somewhat stalled on this one issue, was proceeding as planned elsewhere. And to confound the skeptics, the pattern of annual summits was continued; President Nixon paid a state visit to Russia in June 1974.

It would have been unreasonable to expect the visit to yield important results. By now it was clear, even to Moscow, that Nixon's days in the presidency were numbered. Indeed his trip, apart from its stated purpose, was also prompted by the President's need to gain a respite from his grueling ordeal at home. Also he may have wished to demonstrate that whatever his peccadilloes in domestic policy, he was still indispensable for steering the republic's course in foreign relations. This theme rather unsubtly intruded in Nixon's toast to Brezhnev when he stated that the agreements reached between the two countries "were made possible because of a personal relationship that was established between the General Secretary and the President of the United States." Nixon's reception in the USSR could not be faulted for courtesy and even cordiality.

33. *Ibid.*, p. 135.

Once again he addressed the Soviet people on TV, and (in a gesture of special consideration) was entertained by Brezhnev at his villa in the Crimea. Yet neither he nor his hosts could forget the fact that while all such honors and festivities were taking place, the Judiciary Committee of the House of Representatives was preparing a bill of impeachment.

Under the circumstances the Soviets would not and the Americans could not engage in new far-reaching initiatives and binding engagements. The two agreements signed in Moscow during the visit, one limiting underground tests and the other reducing the respective ABM sites, had been agreed upon in advance of the summit. On the urgent business, such as the Middle East and the situation in Vietnam, where the Paris agreements were constantly being violated, both sides confined themselves in their joint communiqué to pious platitudes.

Nixon's resignation, which duly followed in August, could not affect the Soviets' now changed view of detente. The new administration in Washington, while no longer under siege, was obviously interim; in view both of his personality and the circumstances of his coming to office, the new President was hardly in a position to provide forceful leadership. The oil crisis had affected the Kremlin's assessment of American power; Watergate and its sequels its assessment of the American executive's ability to act decisively and with full Congressional approval in foreign affairs. As mentioned before, the Soviets would still exhibit moderation and caution in areas and on issues about which public sentiment in the US was so strong that no matter how debilitated the US government had become it would still be able and constrained to counter Soviet moves. But in their current mood the Americans would for some time be more concerned with putting restraints on their allegedly "imperial presidency" than on Soviet imperialism.

Paradoxically, while American curiosity and watchfulness over what the USSR was doing in the world at large had lessened, it continued unabated insofar as Soviet domestic affairs were concerned. It ran against the grain for the Kremlin to

countenance American interference and protests on matters as
sensitive as the official treatment of dissidents, whether and
in what numbers Soviet Jews would be allowed to emigrate,
etc. But if the national and ideological pride of some Soviet
leaders was undoubtedly ruffled by this Yankee prying, their
more sophisticated colleagues probably argued that it was better to humor the Americans on such relatively unimportant
issues than to risk militant anti-Soviet attitudes again gaining
the upper hand in the United States. And it was undoubtedly
being pointed out in Kremlin circles that exasperating as were
those foreign pressures concerning what was soon to become
known as human rights, there existed also sound domestic
reasons for appeasing them, up to a certain point, to be sure.
It would have been impractical in any case to deal with bud-
ding dissent in the simple efficacious way this was done under Stalin. Better to exile a Solzhenitsyn rather than surround
him with an aura of martyrdom, and face not only those tedi-
ous protests abroad, but also unfavorable reactions at home.
Similar considerations dictated the desirability of allowing a
sizeable number of Jews who felt that there was no future for
them in the fatherland of socialism to emigrate. Yet the regime
could not accept the proposition that expulsion should be the
only penalty for dissidence, or that any Jew who felt discrim-
inated against was free to leave. To do so, the Kremlin believed, threatened to turn the trickle of dissent into a torrent,
and might lead to a mass exodus of other ethnic groups, as
well as Jews. And so no amount of foreign pressure was allowed to make the government abandon its more traditional
ways of dealing with malcontents, ranging from imprison-
ment and incarceration in psychiatric institutions to barring
the culprit from exercising his chosen profession.

The Jewish problem has for long been a knotty one, for the
Communist regime cannot admit that any ethnic group in its
domains is subject to discrimination or that there could be any
objective reasons for its members to reject the Soviet way of
life. Those who do want to leave, the official line must hold,
do so *not* because of their national origin or even religious
preferences, but because they have fallen under the spell of a

reactionary, xenophobic ideology—Zionism—and thus are in that respect unrepresentative of the bulk of Soviet Jewry.

Ostensibly, at least, the substantial increase in the number of Jews allowed to leave the USSR coincided with the beginning of detente: over 60,000 in 1972–73 against just over 4000 for the entire period 1968–70. And to some extent the numbers of emigres during subsequent years paralleled the ups and downs in Soviet-American relations. But as with the dissidents so with the Jews: the Kremlin never conceded their right to free and unhampered emigration. The number of Jews permitted to leave has always been substantially lower than that of those seeking to do so, and even those eventually allowed to go were often subjected to interminable delays and chicaneries. Jews, and to a lesser degree ethnic Germans desiring to leave the country, were thus used by the regime as a bargaining card in its dealings with the US and West Germany, the Soviet leadership probably congratulating itself on its ability to extract concessions from the West for getting rid of people it did not want to retain in any case. Except for lingering ideological scruples and reluctance to lose valuable human material, some in the Kremlin would very likely opt for expelling en masse the two million or so Soviet Jews. But there are others who must consider that even to appear to yield to foreign pressure on matters affecting Soviet sovereignty is derogatory to national prestige and a potential source of domestic troubles.

The complex ramifications of the Jewish issue were vividly illustrated by the story of the abortive Soviet-American trade treaty. Envisaged as an important element of detente, the treaty included the provision that the USSR should join the states enjoying most-favored-nation status insofar as American customs regulations were concerned. Once in Congress the bill incorporating the treaty became a subject of contention. Its critics, led by Sen. Henry Jackson, insisted on an amendment making the most-favored-nation clause conditional upon the Soviets permitting unrestricted emigration. There ensued a contest involving the White House, the Senate, and the Kremlin. The administration, speaking through Secretary Kissinger, pleaded that the issue should be handled through "quiet diplomacy," rather than through legislative enactments bound to rub raw the Soviets' sensitivities. The Russians made reassuring noises in private discussions, while publicly expressing their indignation over the US Congress presuming to lay down the law for the USSR. A treaty of sorts was reached between the administration and Senator Jackson: he would modify his amendment on the understanding that the Soviets had committed themselves to issue at least 60,000 exit permits per year. The senator then publicly disclosed his victory from the very steps of the White House. Their blood pressure now probably reaching a dangerous point, the Soviet leaders hastened to disavow any commitment on their part; and, to complicate the already strained relations between the State Department and Congress, Moscow chose to reveal that eight days after the Jackson-Kissinger pact, the secretary of state had received a letter from Gromyko warning that "the question . . . is entirely within the competence of our state."[34] The Soviet Union could not countenance any pressure concerning its own citizens, and as a matter of fact, added Gromyko, the number of those within "a certain category of Soviet citizens" who desired to emigrate was falling rather than rising. Still certain ambiguities in Gromyko's letter led optimists on the American side to conclude that this was but a face-saving gesture. Congress went on to pass the bill, but added another amendment, placing a limit of $300 million on credits to be extended to Moscow during the treaty's duration. If the Kremlin had been ready to swallow its injured national pride, it would certainly not do so for a paltry $75 million per year! The Soviets waited for the President to sign the bill into law, then announced shortly thereafter (January 10, 1975) that they would not accept its provisions and that the 1972 trade agreement would not come into force. National honor was saved, and the Soviet leadership had the added satisfaction of throwing another spoke into the wheels of American foreign policy: the White

34. Quoted in Colin Shindler, *Exit Visa* (London, 1978), p. 105.

House blamed its Congressional critics for needlessly endangering detente; they in turn accused the secretary of state of duplicity.

The last phases of the trade drama took place with a new President in Washington. For the Kremlin (as for most of his own countrymen) Gerald Ford was an unknown quantity. Kissinger's continuance in office was from Moscow's perspective a reassuring sign: most parents, even when sorely tried, remain fond of their progeny, and he was among the fathers of detente. But as shown by the story just related, it was obvious that the secretary of state's position had been weakened by the erosion of presidential power and Congress' growing role in foreign affairs. And no matter how brilliant and strenuous Washington's diplomacy, it would not be able to undo the damage to American world standing caused by the events of the last two years.

By the recent rules of "summitry" it should have been Brezhnev's turn to visit the US in 1975. But the new President's advisers were eager to display his ebullient personality to the Soviets, and thus reassure them that the American ship of state was now in firm hands and on a steady course. Since Ford was to visit Japan and South Korea in November 1974, it appeared quite logical and innocent for the Soviets to suggest that the meeting take place nearby in Vladivostok. But someone on the American side should have remembered that the city was in the former Chinese territory forcibly annexed by Imperial Russia as recently as 1860, and its very name—"The Ruler of the East"—had to grate on Beijing's sensitivities.

Strategic arms control was the main subject of discussion at the Vladivostok summit, for the offensive weapons part of SALT I was due to expire in 1977. By now the numerical proportions of the two sides' nuclear arsenals established by the 1972 agreement were being seen by many in the US as having placed this country in a disadvantageous position. But the guidelines laid down for SALT II negotiations at Vladivostok were supposed to allay such fears. Both countries agreed on a common ceiling of 2400 strategic nuclear missiles and heavy bombers, of which not more than 1320 could be equipped with

MIRV mechanisms. On the face of it the agreement when consummated would establish perfect numerical equality between the two superpowers, for the Soviet Union, whose current number of delivery vehicles exceeded the ceiling, would be obliged to dismantle some of them. To a non-expert, the general lines of the agreement appeared eminently fair and a testimony to Soviet reasonability and moderation. But as with SALT I, so in this case, the initial relief in Washington was followed by angry expostulations, the American critics accusing the Russians of trickery and their own leaders of attempting a sleight of hand to conceal yet another acquiescence in Soviet nuclear superiority. The agreement, they claimed, glossed over the all-important question of the disproportion in megatonnage, so that the Soviet nuclear force was allowed to preserve and enhance its greatly superior destructive power. Furthermore, they said, those Soviet missile launchers and bombers that would become inoperative under the agreement were antiquated and slated in any case to be scrapped. For the moment such criticisms on the American side still did not seem to threaten the proposed framework of SALT II. It was only when the Soviets' moves on the international chessboard appeared to violate the *spirit* of detente that its most concrete expression, the strategic arms control process, was jeopardized, and the Kremlin's nuclear capabilities and intentions provided the subject for anxious debates and passionate arguments on this side of the ocean.

To a superficial observer 1975 witnessed a sharp turn in Soviet foreign policy. For the weary Western statesmen who had expected that the 1972 summit would put an end to at least more explicit attempts by the USSR to expand its sphere of influence and destabilize the world scene, the events of the year seemed to demonstrate that once again their hopes had been betrayed: the Soviets were back at their old game, redoubling their "efforts to obtain unilateral advantages" at the expense of the US and its allies, activities they had so solemnly foresworn in the Nixon-Brezhnev declaration of 1972.

Yet a more perceptive analysis ought to have acknowledged that while the Soviets could justly be held responsible for taking advantage of the growing political and economic instability affecting large areas of the world, that instability had been caused in the main by events in the West, or more properly by what the West had been doing or allowed to be done to itself. The overall picture was summarized with admirable candor and succinctness by a North Vietnamese general when he spelled out the reasons that had led his regime to conclude that it could violate the Paris agreements of 1973 and embark upon the conquest of South Vietnam without any fear that the US would retaliate: "The internal contradictions within the US administration and among US political parties had intensified. The Watergate scandal had seriously affected the entire United States. . . . [It] faced economic recession, mounting inflation, serious unemployment and an oil crisis."[35]

No one could or should have expected Moscow to be less sophisticated on such matters than Hanoi, or to forgo piously the opportunities that kept falling into its lap as a result of the West's sins of omission and commission. It was probably genuinely incomprehensible to Brezhnev and his colleagues that anyone could seriously believe detente placed them under the same obligations in 1975 as it had in 1972. The world power configuration had changed. If the US proved unable and/or unwilling to help its friends in various parts of the world, did this mean that the USSR should refrain from assisting its own partisans? To expect such self-denial on the part of the Kremlin would have been as extravagant as to imagine the American government offering its good services to mediate the dispute between Moscow and Beijing!

The new, largely self-imposed limits on American power became evident with the US failure to prevent the fall of South Vietnam and the complete Communist conquest of what had been French Indochina. The South Vietnamese's grudging acceptance of the Paris agreement had been secured by then-President Nixon's solemn pledge to General Thieu that any massive violation of the armistice by Hanoi would be met with immediate American reprisals. But as the South Vietnamese leader should have known, under the new look of American

35. Quoted in New York Times, April 26, 1976.

foreign policy the President could propose, but it was Congress that disposed. At the time no realistic observer could have assumed that (1) granting the state of American public opinion it would be easy to make this country plunge again into the Southeast Asian cauldron; or (2) barring a miraculous upturn in its morale and efficiency the Saigon regime would continue indefinitely to keep at bay its determined and ruthless enemy, which retained military enclaves within the South. Still it could be reasonably hoped that just as the Soviet Union had helped behind the scenes to secure the armistice, so it would use its influence to curb Hanoi's impatience to see all of Indochina under its rule.

The extent to which Moscow was able to control the actions of its North Vietnamese protégés is of course a matter of conjecture. But it is improbable, especially in view of Hanoi's growing estrangement from Communist China, that the North Vietnamese would have launched a major effort to conquer the South had the Soviet Union been categorically opposed to such a move. More likely the Kremlin's reaction to the Vietnamese argument that the overall world situation made it both possible and imperative for them to achieve the goal for which they had been fighting for more than thirty years was on the order of "try it if you will and see what happens."

In January 1975 the regular North Vietnamese forces and the Vietcong launched a major probing operation that met with complete success: conquest of a whole province. Then followed preparations for an all-out effort in the Central Highlands region of the South. This area was seized in March. Hanoi's determination to carry out the enterprise to the end must have been fortified by the fact that Saigon's defeats were turning into a rout, its army abandoning without fight entire provinces, its government incapable of formulating any coherent plan of defense or of effectively controlling its own military and civilian bureacracy. At the same time North Vietnam's assumptions concerning the US reaction were amply justified. President Ford's request for additional appropriations for military aid to Saigon were rebuffed by Congress, which on the contrary passed legislation to terminate all such aid within a

short time. And in view of the country's mood the question of
the US undertaking military reprisals against North Vietnam,
such as were implicit in the pledge which President Nixon
had given to the unfortunate General Thieu, could not even
arise. There remained the melancholy task of evacuating the
last of that once vast body of American personnel: diplomats,
military and intelligence advisers, experts of all kinds who for
more than twenty years had tried to impart viability and
strength to the doomed regime, as well as those thousands of
Vietnamese who felt too compromised by their past roles to
expect leniency from the new masters.

"The historic victory of the Vietnamese people," as the Soviet press dubbed it, was epitomized by that widely dissemi-
nated photograph of the helicopter on the roof of the American embassy ready to lift off and carry away the last
representatives of the world's greatest power, which had
promised to guarantee the independence and security of the
people of South Vietnam. Hanoi's victory was also Moscow's.
At relatively little cost and less risk the USSR enabled a small
Asian country to demonstrate that in the Third World, at least,
the US could not be counted upon to protect its friends,
whether militarily or through other means. As the representative of the Philippine Communist Party was to phrase it at the
Twenty-fifth Party Congress in Moscow: "The confrontation of
socialism with imperialism in Indochina has demonstrated that
with the help of the Soviet Union one can achieve national
liberation without threatening either world peace or detente."[36] For all their caution and inborn conservatism, the
Soviet leaders would have been less than human if some of
that heady feeling had not rubbed off on them.

With its conquest of the country completed as of May 1975,
Hanoi found itself dependent on Moscow even more than before. The Vietnamese Communists never ceased proclaiming
noisily (and it speaks volumes about their appraisal of the US)
that America owed them billions in reparations, which had
allegedly been promised by Washington at the time of the Paris

36. *The Twenty-Fifth Congress of the Communist Party of the Soviet Union*
[stenographic report] (Moscow, 1976), vol. 2, p. 193.

accords. But in realistic terms they had to look mainly to Russia for help in the vast task of reconstructing an economy ravaged by the huge human and material losses incurred both by the South and North. Even more important, with the US out of the picture, it was the People's Republic that the dominant faction among the Vietnamese Communists feared most. Hitherto they had manoeuvred skillfully between the two Communist giants, though moving ever closer to the one giving them tangible material and diplomatic help as against the bracing slogans and encouragements emanating from Beijing. Deeply engraved in the historical consciousness of the Vietnamese people was the memory of the Chinese empire before its decline, constantly asserting and at times imposing its suzerainty over all of the peninsula. By the same token the Vietminh regime—which from its beginning up to the mid 1960s had been so close to Mao's—was now determined to bar Chinese influence in Indochina, and it was already viewing the sizeable Chinese element in Vietnam as a potential fifth column. There was an added bone of contention in Cambodia; within a few weeks of the fall of Saigon, the same drama was repeated in Pnompenh, the difference being that the Cambodian Communists—the Khmer Rouge—were closely linked to Beijing, for reasons similar to those that made the Vietnamese gravitate more and more toward Moscow. Hanoi, with the latter's not-so-discreet encouragement, now made its own bid for the domination of the whole area that once had been French Indochina. The coalition government in Laos was easily brushed off, and the little country became by the end of the year the Laotian People's Democratic Republic, firmly in the Vietnamese-Soviet camp. For a few years the Cambodian Communists were allowed to pursue their violent rule, which in its sheer ferocity and irrationality finds no precedents even in this century. But it was less the inhumanity of the Khmer Rouge—which approached proportions of an attempted genocide against its own people—than its anti-Soviet and anti-Vietnamese position that finally led Vietnam to move to liquidate this last enclave of Chinese influence in the area.

Thus in addition to the damage it did to the image of the

US throughout the world, the fall of South Vietnam brought other solid dividends to the USSR: it now had a militant and battle-tried ally on China's flank. And only a few years before Beijing had been pillorying the Kremlin for not doing enough to assist the heroic Vietnamese people in their struggle against foreign imperialism! But history not infrequently plays such tricks, and a day may come when the USSR will also regret its past commitment to the Vietnamese, and when what has been a brilliant success of Soviet foreign policy will be seen as a dangerous liability.

But since they are not much given to historico-philosophical musings, the Soviet leaders probably saw the last chapter of the Vietnamese saga as an incentive for redoubling their efforts in other parts of the world that were afflicted by political instability and civil wars. Nineteen seventy-four had been a bumper year for political unrest, and now its fruits were ripening. The attempted right-wing Greek coup in Cyprus during the preceding year had destabilized the situation in the eastern Mediterranean and placed in jeopardy the southern flank of NATO. The immediate consequence of the coup was the Turkish invasion and occupation of a large part of the island. This in turn precipitated the collapse of the military junta regime in Athens. Again the Soviet Union was in a position to, and in fact did, exploit the crisis, while being able to claim, and in this case quite correctly, that it had done nothing to provoke it. The initial elation in Greece at the restoration of democratic government soon gave way to angry demands that the United States should compel Turkey to pull its troops out of the predominantly Greek island. American pressure on Turkey was countered by the Soviets' discreet support of it. Then, as Ankara showed no sign of evacuating the island or even of confining its troops to that part of it inhabited by the Turkish minority, the Soviets hastened to express sympathy for Athens and call for an international conference to debate—and in the process very likely to exacerbate—the dispute. The US Congress suspended American military aid to Turkey. This did not stop the Greeks' complaints about Washington's policy— why did the US not *compel* the Turks to pull out of the island?—but had the predictable effect of making Turkey, once the most staunchly anti-Communist among NATO European nations, more amenable to the Soviets' advances. It was an exhilarating game: as each new crack opened in the Western alliance, Moscow kept pounding at it relentlessly and with impunity. Having foresworn the notion of spheres of influence (but in fact recognizing the Soviet one in Eastern Europe) the United States could not remonstrate with Moscow that Cyprus was an intra-NATO affair in which the Russians had no business interfering. Even feeble hints to that effect were met by indignant retorts: the USSR was a great power and could not ignore the danger to world peace inherent in the Greek-Turkish dispute.

John F. Kennedy had had the melancholy insight, as he fretfully complained to Khrushchev in Vienna, that the Soviet maxim seemed to be "What is mine is mine, what is yours is negotiable." But though many years had passed since then, the American policy-makers were still at a loss as to how to deal with that disturbing characteristic of the Kremlin's philosophy of international relations. As Vietnam had already demonstrated, a large segment of US public opinion was chary of extending any tangible help to a friendly nation threatened with internal or external subversion unless the regime in question could be certified as endowed with irreproachable democratic virtues; and, alas, few states, especially in the Third World, could qualify as such. The Soviets on the contrary displayed no qualms on this count: they wooed impartially governments of the most diverse ideological hues, so long as these fitted into their schemes—from a feudal one like that of the King of Afghanistan (as long as he seemed firmly ensconced on his throne) to various revolutionary and radical juntas, a type that happened to be the rule in the new nations formed in the wake of the collapse of the old Western empires.

As a result, US policy was bound to operate under a number of disadvantages in the former colonial areas. Their rulers often saw the US as a friend of their former European masters (even though US pressure had often speeded up the emancipation of former dependencies). More important, the masters of the

new nations found themselves facing economic, ethnic, and
political problems that seldom allowed them, even had they
wished it, to practice democratic virtues. To align themselves
with the United States meant being subjected to constant pressure, if not from Washington then from other influential circles in American politics, to allow free elections and other
paraphernalia of democratic society and to settle amicably,
rather than with arms, their frequent squabbles with their
neighbors. The USSR on the contrary overlooked such imper-
fections and viewed them with avuncular tolerance, provided
that the new state gave promise of adopting a "progressive"
(i.e., anti-Western) orientation. Moscow joined with much of
world (including American) public opinion in condemning the
US for its share in overthrowing President Allende's left-wing
regime in Chile. But the fact that the Iraqi government was
suppressing, and occasionally executing, the country's Communists did not prevent the USSR from continuing its close
ties with Baghdad. And in 1975 Prime Minister Kosygin paid
an official visit to Libya and concluded an agreement that included provisions for the most massive sale of Soviet military
hardware yet to a Third World country. Colonel Khaddafi's
authoritarian and Moslem fundamentalist ways, his strong anti-
Communist views and even erstwhile vituperation against the
Soviet Union were in the Kremlin's eyes of lesser consequence
than his now definitely anti-US tilt and his willingness to re-
plenish the Soviets' hard-currency reserves.

There was little that was new in the general thrust of Soviet
policy as described above. Already under Khrushchev the
USSR had begun to further vigorously its influence in the Third
World. No one could have supposed that detente would put a
stop to such efforts, even though it was hoped on the American side that these would now be less explicitly anti-Western
in nature. But in general US-Soviet competition was a fact of
international life. Never at a loss in justifying their policies,
the Soviets could and did claim that the Americans tended to
blame them unfairly for everything that was not to their liking
in international affairs. It was not Moscow's fault that most
Arab states considered the US definitely committed to Israel

and hence tended to gravitate towards the USSR. It was equally natural that national liberation movements throughout the world saw America as a neo-imperialist power trying to impose capitalist ways and economic domination on the new nations that had just cast off their colonial chains. By contrast the Soviet Union stood ready to offer the new nations its "selfless help" and thus free them from dependence on the international conglomerates and bankers.

In their more candid moods the Soviet leaders and their spokesmen were willing to concede that the Soviet Union was zestfully pursuing power politics. But were the Americans too blind or hypocritical not to see that they too were playing the same game? Did they not chip systematically at Egypt's ties with Moscow, slyly encourage Romania to loosen its ties to Russia, and in general try to sow discontent and dissent within the camp of socialism through agencies such as Radio Free Europe, not to mention the CIA? And as for China . . . It was presumptuous on the part of the US to assume that it had the unique dispensation to interfere everywhere in the world but that for the USSR to do the same was somehow improper and immoral.

Such arguments were now propagated by a growing body of Soviet experts and propagandists centered around several institutes designed to study the West and influence public opinion there. Whereas traditionally the Soviet propaganda effort had been targeted at left-wing and radical circles abroad, and partook of the conspiratorial and revolutionary techniques of the old Third International, Moscow's new-style public relations campaign aimed at influencing "respectable" circles abroad: businessmen, intellectuals, the media community. Figures such as George Arbatov, head of the Institute for the Study of the United States and Canada, were a new breed of Soviet spokesmen, equally at home at an East-West scholarly conference, at their institutes in Moscow, or addressing an American television audience, clarifying listeners' alleged misconceptions about the USSR and letting them in on the essentially peaceful and constructive intentions of the Politburo. As against the stern, occasionally ominous style of official Soviet diplomacy, those unofficial emissaries of the Kremlin spoke the language of international conciliation and understanding, almost entirely free from the Communist ideological idiom. The main gist of their argument was to stress the essential identity of interests between the two superpowers: they shared the main responsibility for world peace and should not allow themselves to be distracted from the path of detente no matter what happened in the Middle East or the Horn of Africa. Quite apart from the supreme necessity of avoiding anything that might lead to the unspeakable horror of a nuclear conflict, the two superpowers had also a common interest in enhancing their commercial and cultural intercourse, both for its own sake and for the beneficent influence it had in promoting friendship between the two nations.

Whatever one thinks of the motivation and sincerity behind such Soviet arguments, they could not offset the new strains which 1975 brought to East-West relations. Perhaps the Americans would not have grown so disillusioned with detente had the Soviets confined themselves to just their *traditional* ways of recruiting friends and expanding influence throughout the world. But that year witnessed the introduction of a new technique of expansion: direct use of military force to establish a fresh enclave of Soviet influence.

To most outsiders, not merely to Washington Kremlinologists, the massive use of Cuban troops in Angola, first to install then to maintain a pro-Soviet faction in power, had to mean an escalation in both the methods and intensity of the Soviet expansionist push in the Third World. For long the Soviets had assumed the prescriptive right to use military force when dealing with what they considered untoward developments within their Eastern European sphere of interest. Occasionally, as in the case of Egypt at the beginning of the decade, they had "lent" their military experts and specialists in sizeable numbers to a sovereign government that had requested them. The intervention by the Chinese "volunteers" in the Korean War, though undoubtedly done with Moscow's approval and to pull its Korean chestnuts out of the fire, was still primarily the result of a decision by Beijing. But Angola was the first case when massive Soviet power—though in the form of surrogate troops—was injected to decide a civil war in favor of the pro-Soviet contender. In the eyes of Washington it was a breach of the unwritten code governing the rules of competition between East and West in the Third World, a violation which made a mockery of the Nixon-Brezhnev declaration of 1972.

For their own part, the Soviets could and did claim that the dissolution of the Western empires in Africa created an entirely new situation in which the code of detente did not place any special constraints on them. Again, they went on, the West claimed some special prerogatives for itself. It assumed that once European rule collapsed in various parts of Africa it could by right be replaced only by American economic and political influence and that the USSR should keep out. There was a plausible element in the Kremlin's argument: both superpowers, as well as China, had for some time anticipated the inevitable collapse of the decrepit Portuguese empire, and insofar as its most important part, Angola, was concerned, each had its favorite among the variety of guerrilla factions fighting against Lisbon's ever more precarious hold over the country. In the early 1960s, the CIA was already discreetly helping, mostly with money, the then most prominent of them, the National Front for the Liberation of Angola (FNLA); but American support could be neither extensive nor publicized in view of the US NATO links with Portugal. The Soviet entrant in the race was the Popular Liberation Movement (MPLA). At a later stage Beijing expressed its sympathy with but provided little tangible help to a group called UNITA. As long as the armed struggle against the Portuguese armies continued, the main resistance movements preserved a semblance of unity. With the French and British empires in Africa disbanded, it could be only a question of time until little Portugal, the West's poorest country and expending the lion's share of its budget on the colonial wars, had to give up the struggle. As of 1964 a number of African and Arab countries had recognized the insurgent Angolan "government," the FNLA's leading role in the struggle being acknowledged by the fact that its leader,

Holden Roberto, was proclaimed the head of the provisional regime.

The USSR could not be pleased with the prospect of independent Angola joining the pro-Western group of African states. Already in 1968 Soviet African experts made clear their antipathy towards the FNLA, in a rather rare instance of Soviet criticism of any national liberation movement prior to its victory over its country's imperialist masters: "Holden Roberto and his entourage pay homage to 'Western freedom,' display servility toward the US and indulge in anti-Communist rhetoric. . . . He slanders the Communists and patriotic forces of Angola. . . ."[37]

In 1974 the authoritarian regime of Portugal was overthrown by a coup inspired and led by the military, frustrated by the ruinous and futile struggle to hold on to the colonial possessions. The new democratic regime, which included several parties, among them the Communist, decided immediately to authorize self-determination by its African territories. For a while there arose a possibility of Portugal itself joining the Communist camp, for its very pro-Moscow party had considerable support among the younger elements of the officer corps. But while the Communists' peaceful and not so peaceful efforts to take over power in Lisbon were ultimately frustrated, there can be little doubt that Portugal's left-of-center regime proved on the whole favorable to the MPLA cause in Angola.

Though the three main Angolan nationalist movements agreed at first to form a coalition government, no one familiar with the local scene could believe that the compromise would endure. There had been no rudiments of representative institutions in the country; each of the contending factions had an essentially tribal base, even though the MPLA counted sympathizers among the local Portuguese and mixed-blood intelligentsia, which as in most colonial areas had imbibed heavily of the Marxist-Leninist idiom. There were numerous clashes, and by April 1975 a full-fledged civil war had begun. The

37. Y. S. Ogisian, *The National Revolution in Angola, 1961–1964* (Moscow, 1968), p. 167.

FNLA and UNITA combined their forces against the pro-Moscow faction. The latter, at first at a disadvantage, soon found itself the recipient of massive amounts of military materiel from the Soviet Union, including MIG jets, ground-to-air missiles, and other sophisticated weaponry. As for the FNLA and UNITA, they were furnished some weapons, mostly antiquated and of Chinese and West European manufacture, by Zaire. But the decisive factor proved to be the infusion of Cuban troops, which began to arrive in July 1975. The first small number—mostly specialists sent to instruct the MPLA in handling the Soviet weapons—were soon joined by regular army formations. The total number was estimated at about 3000 in November, and it grew to 12,000 by the following February. Hardly an impressive number, but within the context of a struggle fought by primitively trained and armed guerrilla forces, the Cubans (themselves instructed by the Soviets) made the tide of battle shift decisively in favor of the MPLA. There was a temporary setback to the latter's fortunes as US military supplies began to flow to the FLNA forces and South Africa intervened with its own troops on the anti-MPLA side in the last months of 1975. But there arose a general outcry at this intervention by the racist South African regime, and its troops were withdrawn in January–March 1976.[38] The Cubans stayed. On February 11, 1976, the Organization of African Unity, comprising most of the independent states of the continent, recognized the MPLA regime as the legitimate government of Angola.

There have been, even among some American circles, various attempts to rationalize the Soviet-Cuban intervention in the affairs of an African state whose internal politics could not in the slightest affect the security of the USSR or of the Soviet bloc. There was a half-hearted attempt to portray the military aspect of intervention as an entirely Cuban operation. Equally unconvincingly, it was protested that the Cuban troops entered the scene only following the South African attempt to smash the Angolan progressive elements. Yet even if one assumes that Castro welcomed the chance for Cuba to become

38. Colin Legum, *After Angola* (New York, 1976), p. 38.

the vanguard of the advance of Marxism-Leninism, not only in Latin America but throughout Africa, the burden of logistics as well as of financing the Cubans' incursion had to be borne by Moscow. And, as we have seen, Cuban military units entered the country's civil war long before South Africa's.

What must be of interest to us, rather than the complex story of Angola's civil strife, which as of this writing still continues, is a reconstruction of the Kremlin's motives and expectations in undertaking this bold new venture. There are some indications that the move was made with full awareness of establishing a new threshold of Soviet expansion, and that Moscow watched closely for foreign, especially American, reactions to this innovationist technique. (The Organization of African Unity, then headed by the notorious Idi Amin of Uganda, at first tended to condemn the Soviet action as a gross interference by an outside power in the affairs of Africa, but then, as we have seen, acquiesced in it.) There were cogent reasons for employing Cuban rather than Soviet soldiers; the first was the fiction that another Third World power was offering its military support to the progressive cause in Africa. Second, a large part of the Cuban contingent was black; and third, Spanish could be understood by the Portuguese-speaking Angolan elite. And at least some in the Kremlin must have thought it only fair that Castro should repay some of the largesse and protection lavished on him by the Russians.

As to the American reaction: since Cuban combat troops were introduced gradually and at first in small numbers, the experiment could have been terminated under various pretexts ("The Cubans were no longer needed once the South Africans had pulled out," etc.) had Washington unexpectedly chosen to take forceful countermeasures. But the prevailing US attitude toward any involvement in a distant area was still epitomized by the cry: "No more Vietnams." You begin, argued Congressional sources, by sending military supplies. You end up by having to send half a million American boys. President Ford's and Kissinger's pleas for substantial funds for arms to help the anti-MPLA forces were decisively rebuffed by Congress, which then passed specific legislation barring such help.

And so Soviet influence became permanent in a strategically vital African country. With its extensive sea coast and its common borders with Zaire, Zambia, and Namibia (whose Marxist-oriented guerrilla movement has been fighting South African rule), Angola provides an important Soviet bridgehead in the highly unstable and easily inflammable political scene of southern Africa.

The most important lesson and corollary of the Angolan crisis involved the Soviets' ability to secure the world's acquiescence in their new technique of expansion in an area thousands of miles from any region of direct importance to the USSR or its allies. One who knew the Kremlin's way of operating could easily predict that, seeing the experiment succeed, it would try to duplicate it elsewhere. And it is a melancholy fact that there are few countries in sub-Saharan Africa where the introduction of a few thousand well-trained troops could not overthrow the existing regime or tilt the scales of a civil war in favor of the faction the USSR has chosen to support through Cuban or other surrogate troops. Southern Africa, with its fragile regimes and its mineral wealth, remains a tempting prize, and much as the new African nations may resent foreign intrusion, the Soviets, especially when wearing the Cuban mask, do not usually arouse the emotions associated with their past colonial masters.

Yet the apparent ease of the Angolan enterprise concealed its potentially very serious consequences to the Kremlin. American public opinion might not have sanctioned doing anything about the Soviets' Cuban gambit, but for many on this side of the ocean, both within and outside the government, it stripped detente of much of its previously hoped-for meaning. Few Americans had indeed considered whether what the West had done to itself since 1972 was responsible for what appeared to them as unprovoked and brazen Soviet conduct. American remonstrations with the Kremlin were met by the kind of rejoinders with which we have already become familiar: what did Angola have to do with Soviet-American relations? When Secretary Kissinger on one of his visits to Moscow mentioned to newspapermen that he proposed to dis-

cuss the Cuban presence in Angola, the Kremlin's reaction was
one of mockery: Dr. Kissinger could indeed discuss the issue,
but with his own advisers, since the USSR could not interfere
in matters pertaining to Cuba's sovereign rights.

As originally conceived, detente had, one might say, three
main ingredients: restraint, reciprocity, and rhetoric. As 1975
drew to a close it was clear that the element of reciprocity had
definitely lapsed: on the chessboard of international politics,
the USSR continued the pursuit of "unilateral advantages" that
it had supposedly foresworn in 1972. It could still be arged
that the two powers adhered to the maxim of restraint insofar
as they continued to observe the provisions of SALT I, even
though some American experts were already arguing that
Moscow was taking advantage of every conceivable loophole
in the agreement. Certainly, on the political side, Angola could
hardly be held as an exemplary case of restraint by one superpower, while the Soviets, never at a loss in an argument,
claimed that Washington's strenuous wooing of Egypt and
China was hardly in line with the spirit of Soviet-American
rapprochement.

For the moment rhetoric remained the strongest remaining
element of detente. Mutual recriminations and accusations of
bad faith still continued, but at a conspicuously more subdued
level than during the Cold War or the more strained periods
of "peaceful coexistence" (the not precisely accurate description of relations between the two superpowers between 1956
and 1972). In their domestic and foreign policies the Soviet
leaders have never underestimated the power of words and
symbols. They were quite aware of the fact that for the people
of the world, including their own, "detente" acquired a meaning quite apart from the complex of agreements and under-
standings that constituted its substance: it also became an
expression of hope that no matter how strained East-West relations might become, both sides still adhered to their determination to settle their differences through dialogue rather than
nuclear arms. This symbolic propaganda value of the term
made the Soviet leaders stress continuously that while "some
circles in the West," or even the administration in Washing-

ton, might be scheming to return to the Cold War, the USSR would continue to adhere to the spirit of detente.

The rhetorical side of detente was capped by the Helsinki Final Act, the final consummation of the Conference on Security and Cooperation in Europe. On its face the act was the charter of European peace and comity, a solemn pledge by thirty-four continental states, from both East and West, as well as the US and Canada, to abide by enlightened standards of international conduct as well as to observe in their domestic affairs certain rules consistent with human rights. Ironically, just as much of the substantive content of detente was in fact being eroded, its general principles were reaffirmed and extended insofar as the whole European community was concerned.

The irony of the situation was enhanced by the fact that it had been the USSR that sponsored the idea of a general European pact; yet already at its signing in August 1975 there were clear indications that the document was bound to prove deeply embarrassing to the Soviets, diplomatically and in terms of propaganda.

The notion of such a conference and agreement had been pushed by the Soviets for many years, originally for quite understandable reasons. Long frustrated by its inability to get a *formal* recognition of the post–World War II settlement and the frontiers established "provisionally" in Potsdam in 1945, the Kremlin had sought another way of obtaining international sanction for the European status quo—the division of Germany, Poland's western frontiers, the Soviets' own territorial acquisitions, etc. The series of agreements West Germany concluded with the USSR and Poland between 1969 and 1972 appeared to remove the need for a general European conference to affirm the principle of the inviolability of European frontiers and other features of the new continental order. But still the Soviets pushed for a European security conference. The enterprise finally flowered into a document containing a variety of provisions pledging the signatories not only to cooperation insofar as European security was concerned but to a whole range of economic, scientific, and humanitarian mea-

sures, many strikingly different from the actual practices prevailing in some of the participating states. A visitor from another planet suddenly transposed to Helsinki at the final stage
of the conference would have undoubtedly concluded that here
was a great charter of peace and freedom bound to secure for
decades to come not only Europe's peace, but its orderly progress toward democracy and prosperity.

The original, i.e., Soviet, motives in promoting Helsinki were
expressed by the document's statement that "the participating
states regard as inviolable each other's frontiers, as well as the
frontiers of all states in Europe," accompanied by the usual
abjurations of the use of force or intimidation and of encroach-
ments on others' sovereign rights, and by pledges of non-interference in their internal affairs, etc. "Basket Two" of the act
listed various measures of economic, scientific, and cultural
cooperation which this new supposed concert of Europe
promised to implement. Many of them were simply common
sense, already in practice and of little if any political significance. The Soviets had for some time sought to widen professional and trade ties with the West, but this was already taking place quite independently of Helsinki.

The famous "Basket Three" of the Helsinki Final Act has
since surfaced as a major source of embarrassment to the Soviet government. But in signing its provisions the Kremlin had
little reason to assume that it was thus opening itself up to
what in its eyes has constituted foreign prying in Russian domestic affairs as well as providing at least a verbal tool of protest to dissidents within its own society. The USSR after all
had already been a signatory to a number of international
agreements and declarations whose occasional violations by
Moscow had been accepted by the international community
as ineluctable facts of life.

On their face the provisions of Basket Three were unsensational. They required the signatories not to place undue obstacles in the path of citizens seeking to be reunited with their
families abroad or to marry foreigners, "to ease regulations
concerning movement of citizens from other participating

states in their territory," and to allow members of religious and professional organizations resident in one of the states to maintain contacts and hold meetings with their fellow believers and professional colleagues in foreign parts. To abide literally by such promises would have been strikingly in contrast with the accepted Soviet practices in such matters, and it is difficult to conceive that the Russians thought that anyone could seriously expect them to do so.

Contrary to what is generally believed, the main thrust of the Helsinki agreement concerning human rights is found, however, not in "Basket Three," but in the introductory declaration ("Basket One") which stated plainly that "the participating states will respect human rights and fundamental freedoms, including freedom of thought, conscience, religion. . . . . They will promote and encourage the effective exercise of civil, political, economic, social, cultural and other rights . . . [which] derive from the inherent dignity of the human person." [39] Well, most of those admirable principles are enshrined in the text of the Soviet constitution. Yet few would argue that such rights have in fact been respected by the Soviet government, or that even partial observance of them is compatible with an authoritarian regime such as those of the USSR or its phalanx of Eastern European protégés such as Czechoslovakia. In simplest terms, the Soviet Union and its Communist friends solemnly undertook as their international obligation to do something neither they nor any non-Communist signatory to the Helsinki Act could seriously believe they would do!

The solemn setting of the signing of the Helsinki Final Act, with each country represented by its head of state or government, including such potentates as Ford, Brezhnev, and the aged Marshal Tito, served only to enhance the surrealistic nature of the occasion. Had the noble principles enunciated in the agreement been put in effect there would be little point in continuing our narrative, or dealing at length with such trou-

39. Quoted in *Keesing's Contemporary Archives for 1975* (London, 1976), p. 27302.

blesome subjects as nuclear weapons, NATO, the Warsaw Pact, etc. "And they lived happily after" would be a fitting conclusion at this point!

Their adherence to what they must have believed was a harmless piece of rhetoric was to prove for the Soviets a not unserious psychological error. The Helsinki process continued through regularly scheduled international conferences, which as might be expected usually deteriorated into unseemly squabbles, with the West raising the issues of Jewish emigration, the more notable cases of persecution of the Soviet dissidents, etc., and the Russians grasping at some cases of racial discrimination in the US, British treatment of the IRA prisoners in Northern Ireland, etc. But there can be no question that the problem of political persecution in the USSR has become, so to speak, internationalized and that foreign governments have now acquired a legal handle for their complaints about the Kremlin's treatment of its citizens. It has been a matter of considerable contention whether foreign involvement (especially that coming from a government) in the question of human rights in the Soviet Union is desirable and/or effective, and we shall have occasion to return to this difficult problem.

As with detente as a whole, it would be an oversimplification to consider Soviet accession to the Helsinki accords as pure cynicism. Quite uninhibited in the liberties they have taken with their international obligations, and scornful of agreements not backed by tangible sanctions, the Soviet leaders have at the same time, and incongruously, sought to have their country recognized as a respectable and peaceful member of the European community. And so Helsinki was not entirely a monument of futility, but also, even if in a very small and tentative way, a monument to hope, and as such it epitomized the fate of detente as of the end of 1975.

4

China's Lengthening Shadow

As it entered the twelfth year of its rule, the Brezhnev regime could boast of having achieved domestic stability unprecedented in the history of the Soviet Union. Stalin had ruled through terror and purges of the state and party apparatus, which like violent spasms periodically gripped the ruling oligarchy. The politics of the Khrushchev period, though infinitely more humane, still witnessed intermittent and violent clashes within the ruling oligarchy, the last of which ushered in the Brezhnev era. Whatever conflicts arose within the Soviet elite would now be settled decorously in the privacy of the Politburo. Some of its members, presumably mainly those who opposed or criticized the General Secretary's leadership, were dismissed or demoted, but in each case without the controversy and publicity that characterized Khrushchev's encounter with the "anti-Party group" in 1957 or his sporadic shakeups of the Soviet administrative system. A minister or a local party secretary could now be reasonably sure that he would not be dismissed or disgraced because of a sudden whim or irritation on the part of the leader. Examination of the past crimes and follies associated with Stalinism had now almost

entirely ceased, and many older and middle-aged members of
the ruling group could breathe more freely. Dissent, though
not eliminated, was contained by selective chicanery rather
than outright terror: expelling some troublemakers, imprisoning others, and eschewing those sporadic fits of liberalism that
had allowed the country and the world to learn about Solzhenitsyn, Sakharov, etc.

Economically, though diverting enormous resources (var-
iously estimated at 14–20 percent of the GNP) to armaments,
the regime had been able to assure a steady (if modest by comparison with that of the 1950s) economic growth and improve
the lot of the consumer. The rise in the standard of living, if
not impressive (the Soviet Union was in this respect behind
even the other Communist countries in Europe), was steady.
Having scrapped the Khrushchev plans of economic reorga-
nization, the government experimented at first with some
schemes of industrial decentralization of its own. But the inherent conservatism of the ruling elite soon reasserted itself,
and industrial management reverted to its traditional pattern.

A future historian assessing the Brezhnev era may well have
reason to conclude that it persisted in ignoring or sweeping
under the rug the most perplexing problems of Soviet society
and of world Communism, but as of 1976 one could not begrudge it credit for a large measure of success in the management of its domestic as well as foreign affairs. And from the
point of view of the Soviet ruling class, the main yardstick of
this success was a comparison with what was happening in
other major areas of the world. "It is precisely during the past
[five] years that the capitalist world has experienced an economic crisis, the seriousness and depth of which . . . can only
be compared with the crisis at the beginning of the 1930s,"
said Brezhnev at the Twenty-fifth Party Congress in February
1976.[1] And then in a not-too-veiled allusion to the recent developments in the US, the General Secretary noted the "intensification of the ideological and political crisis of bourgeois
society." Referring unmistakably to the misfortunes that had
befallen his erstwhile partner in detente, Richard Nixon,

1. L. I. Brezhnev, *Following Lenin's Path*, vol. 5 (Moscow, 1976), p. 497.

Brezhnev proclaimed somewhat smugly that this crisis had afflicted "all the institutions [of the capitalist system], the bourgeois political parties, etc. It has undermined the most basic moral standards, with corruption becoming ever more open and reaching into the highest areas of the state. Spiritually and culturally [capitalist society] continues to decline while crime and delinquency are becoming more rampant." All in all, concluded the Soviet leader to the loud cheers of his audience, "there is no future for capitalist society." [2]

Yet while they cheered, at least some of the delegates to the Congress must have reflected that many of the social and spiritual ills attributed to the capitalist world had counterparts in their own society. To be convincing, the claim that the Soviet system was superior to that of the West had to be based on more tangible evidence than allegations about the latter's moral and ideological decadence and economic decline. (Did not Stalin's crimes, excoriated not so many years before from the very same tribune by Brezhnev's predecessor, dwarf those of the capitalist politicians? Was the Soviet Union, which in 1975 had experienced the worst harvest in more than a decade, free from economic troubles?) And so it was in talking about foreign policy that Brezhnev was most persuasive in his claim that the authority and might of the USSR had grown while that of the US and its allies declined: "The Soviet people are proud of having given considerable help to Vietnam in its struggle against the imperialist intruders. . . . Laos and Cambodia have followed Vietnam in winning their freedom . . . the sovereignty of the German Democratic Republic has now been universally acknowledged [despite] the efforts of American imperialism . . . ; socialism has sunk its roots in Cuban soil." [3] And there followed other examples of the Soviets' successes and of the "imperialists' " discomfitures: Angola and other former Portuguese colonies in Africa whose liberation, Brezhnev implied, was made possible by Soviet help; the especially close ties that had been established between the USSR and India; etc. And "talking about our relations with Asian

2. *Ibid.*, p. 480.
3. *Ibid.*, p. 454.

states, it is necessary to refer to our good neighbor, Afghanistan, with which not long ago we concluded an agreement
extending our by now almost half-century-old Treaty of Nonaggression and Neutrality."[4]

In view of the preceding it is understandable why, in speaking of relations with the US, Brezhnev's tone was rather
haughty: yes, detente had proved of great importance and been
of great benefit not only to both states, but to the world at
large "in lessening the danger of a new world war and in
strengthening peace." Unfortunately there were "influential
forces" in America that tried to sabotage good relations between the two countries and called cynically for a new armaments race by the US and NATO. These same forces had attempted to use trade as a lever for interfering in the internal
affairs of the Soviet Union. And "it is hardly a secret that there
are special complications connected with those aspects of
Washington's policies that threaten the freedom and independence of various nations. . . . We have opposed such activities in the past and so we shall in the future."[5]

There was only one dissonant note struck in this otherwise
self-confident recital of Soviet achievements and condescending account of the West's mounting difficulties and futile
strivings. In the part of his foreign policy report dealing with
China, Brezhnev's tone revealed both nervousness and exasperation. Beijing, in his words, "has been trying feverishly to
destroy detente, to sabotage any disarmament efforts, and to
sow distrust and hostility between various states. Its basic aim
is to provoke a world war, and to exploit it for its own purposes." But after this horrendous indictment, the General Secretary, or his speechwriter, was capable only of a rather lame
conclusion: "This policy of Beijing's is deeply at variance with
the interests of all nations."[6] And instead of threats directed
at the Chinese Communist leaders should they continue such
evil doings, there followed an incongruous appeal: Wouldn't
China (Brezhnev here eschewed the term "People's Republic,"

4. *Ibid.*, p. 464.
5. *Ibid.*, p. 471.
6. *Ibid.*, p. 459.

until recently customary in the Soviets' references to Mao's regime) normalize its relations with Moscow? The latter stood ready to base their mutual relations on the principle of peaceful coexistence. More than that: should the Chinese leaders return to the path of true Marxism-Leninism, the Soviets for their part would be more than willing to resume their previous fraternal cooperation with the People's Republic. Again, many in the room must have silently wondered how a regime whose leaders were capable of the diabolical machinations ascribed to them by Brezhnev could ever be trusted, let alone become a friend of the Soviet Union and a faithful follower of proletarian internationalism.

What *partly* relieved Soviet anxiety about China was the obvious debility of Mao, and the pleasing possibility that with his death the already serious factional strife within the Chinese Communist Party might intensify to the point of throwing the world's most populous nation into complete disarray, perhaps a civil war. One of the factions contending for power might then try to get Moscow's backing; in any case the horrifying prospect of a modernized and industrialized China disposing of a powerful nuclear arsenal would be indefinitely postponed.

It is therefore all the more remarkable, and a testimony to the depth of the Soviets' bitterness and their detestation of their huge neighbor, that when the Great Helmsman did actually depart this world, the Kremlin found itself unable to react in a truly tactful and conciliatory manner that might persuade at least some Chinese that it was ready to let bygones be bygones. Traditionally, a foreign Communist leader's death would be announced in Russia with great solemnity, followed by effusive condolences from the Soviet hierarchy and long panegyrics extolling the deceased. But it was merely a six-line communique on a back page of the September 10, 1976, issue of *Pravda* that informed the public of the death of Mao. Equally curt, to the point of rudeness, was the message addressed on the occasion by the Central Committee of the Communist Party of the USSR to the corresponding body in China (and which the latter contemptuously refused to accept). Neither Brezh-

nev nor other senior leaders saw fit to repair in person to the
Chinese Embassy in Moscow and to inscribe their names
among the mourners, delegating this task to Gromyko and another lesser Politburo member.

Having so graciously acknowledged the death of China's
leader, the Soviets proceeded unabashedly to press their
friendship upon Beijing. A *Pravda* editorial of October 1, 1976,
listed all the past services the Russian Communists had rendered their Chinese comrades, as well as the untiring efforts
of the Soviet government to restore the once so close friendship and collaboration between the two countries. There followed a rather plaintive listing of the most recent steps in that
direction by Moscow. The Soviets, the editorial went on, would
be happy if Beijing acknowledged that the 1950 Sino-Soviet
Pact of Friendship, Alliance and Mutual Help (not due to expire until 1980) was still valid(!). And if the Chinese were not
willing to go that far, wouldn't they at least sign a treaty of
nonaggression with the USSR? How about expanding trade
between the two countries and resuming cultural and sport
contacts?

The Soviet press watched greedily for any signs of internal
unrest in the People's Republic following Mao's death. But the
Kremlin soon had to realize, if it had not known all along, that
no Chinese political leader of stature was likely to trust its
honeyed words. Anti-Soviet—more properly anti-Russian—
feelings were by now deeply ingrained in Chinese Communists of all ideological persuasions. There remained the consolation that while China did not lapse into civil war, the in-
tra-party struggle for power did intensify, the cult of Mao was
visibly receding, and the unsettled state of Chinese politics
helped reveal the country's economic weakness, the enormous
harm caused in all spheres of national life by the Cultural Revolution, and other negative consequences of the late Great
Helmsman's reign. The vicissitudes of what the world had once
assumed to be a monolithic Communist Party and a country
run with an iron hand obviously gave the Soviets a great deal
of satisfaction. Caustic comments in the press accompanied
the news of the fall of the "Gang of Four," i.e., the faction

headed by Mao's widow. Here were the Chinese Communists, once so proud of their supposed ideological purity and scornful of Soviet practices, now freely confessing the errors of the past and introducing many economic and social measures they had once condemned as "revisionist." But no matter who is talking about China, whether the press or the highest leaders, and regardless of the issue under discussion, one always detects a certain tenseness in Soviet comments on the subject.

Like its relations with the US, those with China affected the whole spectrum of the Soviet Union's foreign relations. In a sense the Sino-Soviet antagonism cut deeper into Moscow's world view than problems associated with its tortuous relations with the US. Unlike the latter, China could not be thought of as a potential major military danger—certainly not in 1976 and for many years to come. But the spectacle of a great Communist power irreconcilably hostile to the USSR called into question the whole rationale of Soviet ideology and hence, in a sense, of the Soviet system itself. To be sure, after the Cultural Revolution and especially after Mao's death Beijing was not an effective rival to Moscow in their world-wide competition for the allegiance of other Communist parties and "liberation movements." But the Chinese challenge on the ideological front remained, and once the People's Republic mended its internal politics and economy, its appeal to many of the current Soviet protégés, especially those in the Third World, could once more become quite troublesome.

But even in the midst of its post-Mao troubles, the intractable Communist giant complicated the Kremlin's policies and plans all over the globe. For a long time one of Moscow's cherished schemes had been to enlist Japanese capital and industrial skills in the task of developing Siberia and the Soviet Far East, and eventually to draw Japan out of the American orbit. The first step towards both those aims had to be a Soviet-Japanese peace treaty, by now a formality to be sure, but psychologically very important to Japanese public opinion. But Tokyo insisted that the treaty must return to Japan the four southernmost Kurile Islands, which along with the northern part of the chain and South Sakhalin had been seized by the

Soviets in 1945. Both militarily and as real estate the territory
in question was probably the least important of the Soviet
conquests of World War II. Nevertheless, this was one concession the Kremlin felt it could not afford, even for the sake of
wooing the third largest industrial power in the world. For
many years China had called upon the USSR to acknowledge
in principle that the vast territories ceded in the nineteenth
century by the Manchu Empire to Russia had been obtained
illegitimately. As early as 1954 Mao demanded that Mongolia,
long a Soviet satellite, be returned to Chinese suzerainty. Even
the very minor border dispute between the two Communist
powers became the subject of an interminable wrangle; ever
since 1969 a Soviet deputy foreign minister has gone periodically to Beijing, there to be confronted by impossible Chinese
demands for settling what was in its essence a trivial issue
concerning a few square miles here and there. A territorial
concession in *Asia* would thus set a bad precedent. As Brezhnev phrased it at the Twenty-fifth Congress, "There are people
in Japan, obviously inspired by some outsiders, who in connection with the peace settlement are trying to make completely groundless and illegal claims upon the USSR."[7] Here
was another example of the usually skillful and flexible Soviet
diplomacy outsmarting itself, or more precisely, being driven
by fear and loathing of Beijing into playing exactly into its
hands. For the sake of a few fishing villages (as it had to seem
to the world) Moscow jeopardized its *rapprochement* with Tokyo and made it much easier for the Chinese Communists
themselves to establish closer relations with the Japanese. The
Sino-Japanese Peace and Friendship Treaty of 1978, concluded despite the Soviet Union's repeated warnings to Tokyo, marked one of the most signal failures of Brezhnev's
diplomacy. The signatories joined in condemning "hegemonism," which Moscow correctly and irritably took as
a reference to its own policies. More significantly, the treaty
enhanced the frightening prospect of Japan's enormous industrial and technical resources helping to modernize the economy of the world's most populous state—one which, even if

7. *Ibid.*, p. 471.

its present differences with the USSR should get papered over, will always be feared by the Soviets. It is not a good omen for the two nations, and perhaps for the world at large, that anything having to do with China tends so often to drive the present occupants of the Kremlin into a frenzy. And, needless to say, their feelings are heartily reciprocated by the other side.

The inability to effect any change in Beijing's anti-Soviet posture was not the only major disappointment that 1976 brought to the Kremlin. On March 15 Sadat's Egypt abrogated its 1971 Treaty of Friendship and Collaboration with the Soviet Union. This could hardly have been a surprise: except for a few weeks late in 1973, the whole trend of Soviet-Egyptian relations ever since 1972 had pointed in that direction. The last straw was the Soviets' refusal to reschedule the payment of the very considerable debt owed them by Egypt; Moscow saw little reason to subsidize a regime that was moving closer and closer to the US. But apart from its immediate consequences, the Egyptian fiasco called into question another basic premise of Soviet foreign policy. The case of China had already illustrated how victories of Communism in foreign parts could turn into defeats for the USSR. Sadat's defection seemed to carry the same warning about the Soviets' attempts to expand their influence in the Third World. How much money had the USSR poured into Egypt since Nasser's time! And as recently as October 1973 the Soviets had at least seemed to be ready to risk a confrontation with the United States to save Egypt from a threatened military disaster. But the only results of the long-standing Soviet liaison with Egypt were enhanced instability and danger of war in the Middle East. Was it really in the Soviet national interest that this should be so? Imperial Russia had been authoritarian, to be sure, but public opinion did play a role in shaping foreign policy, and there is little doubt that in a case like this some voices would have been raised saying, enough!—those foreign entanglements in faraway places bring us nothing but discomfitures and dangers. But after sixty years of Soviet rule, the only voice that could be heard was that of the government and its propagandists. And all the threads of foreign policy were in the hands of a handful of oligarchs incapable of changing habits and beliefs that had sustained them for decades: the Soviet Union had to advance on the world stage, thus offering repeated proof of the superiority and vitality of its system and conversely demonstrating the decadence and weakness of the West, whose social system had "no future."

Stymied in its efforts to reach a formal accommodation with China, thwarted repeatedly in its probes for some kind of understanding with the US about the People's Republic, Soviet foreign policy could thus, almost as if by inertia, follow the beaten track: expansion behind the facade of detente. In professing their scrupulous adherence to the latter, the Kremlin leaders sought at the same time to redefine its meaning and scope. Beginning in the mid-1970s, they began to insinuate that detente had little to do with world politics at large but applied primarily to Soviet-West relations. It was thus unsporting of Washington, indeed an attempt to revive the Cold War, to complain about what the Soviets or their Cuban or Vietnamese protégés were doing in Angola, the Horn of Africa, or Cambodia. What did those activities have to do with East-West trade, or negotiations about SALT II? A similar reinterpretation was applied, in this case almost immediately upon its proclamation, to the Helsinki Final Act. The USSR had promised therein (in fact it had always promised) to observe human rights. But no one, by the furthest stretch of their imagination, could therefore deduce that Helsinki gave foreigners the right to pry into the Soviets' domestic affairs, or to expect its government to tolerate antisocial and criminal activities by its own citizens. Had it not been attended by such serious implications, this verbal propaganda battle about detente might have appeared comical: many disgruntled Americans felt that they had been fooled by the very term—this wretched foreign word had made them believe, despite all their previous experience, that this time the Russians would behave, and now look! But how could they be accused of being untrue to detente, argued the Soviets with a straight face: had they not tried their utmost to increase East-West trade, opened their country widely to foreign tourists, sponsored and encouraged all kinds of cultural, scientific, and sporting exchanges and contacts?

To be sure there must have been people within the Soviet hierarchy who felt that what remained of detente was still attended by inconveniences, if not real dangers, both for their own regime and for those of the Communist states of Eastern Europe, perhaps for world Communism at large. Domestically, the influx of foreign tourists was creating additional problems for the security forces of the USSR and those of the "fraternal countries." Certain inhibitions had to be observed in dealing with the dissidents: undoubtedly without detente many more among the most prominent ones would have been made to suffer for their views in jails, psychiatric clinics, or Siberia, rather than being allowed or forced to go abroad. If within the Soviet context such self-imposed constraints did not unduly hamstring the government, in some other Communist regimes the need to display even a semblance of tolerance could, and in one case did, lead to very serious political consequences.

Domestically and in economic terms the impact of detente was most strongly felt in Poland. The Gierek regime, taught by the lesson of the workers' riots in 1970, had sought strenuously to improve the lot of the consumer and to bolster the shaky prestige of Communism in this traditionally Catholic country by promoting rapid economic growth and a rising standard of living. It did so largely on the strength of greatly expanded trade with the West, and, especially after 1972, very substantial credits from the same source. For a few years, 1972–75, the formula seemed to work like magic. Poland's GNP rose during the period by amost 60 percent; real wages increased at an annual average of 8 percent.[8] It should have been clear throughout that this economic boom was being built upon very shaky foundations. Agriculture was neglected in favor of heavy industry. And the latter, financed by foreign loans, was geared largely to producing for the Western market, already shrinking because of recession. In 1975–76 the bubble began to burst. Much of Poland's now reduced hard-currency earnings had to

8. Figures in Adam Bromke, *Poland: The Last Decade* (Oakville, Ont., 1981), p. 67.

be devoted to servicing its foreign debts. And as we shall see in the final chapter, there is also strong evidence that at about the same time the USSR resumed its attempts to exploit economically the "fraternal countries," Poland included. Inflation began to cut into the workers' economic gains. Trying belatedly to restore balance to the economy and to cure its faltering agriculture, the regime in June 1976 announced substantial increases in food prices. As in 1970 this step precipitated serious riots in a number of industrial centers, raising once more the spectre of a mass workers' revolt against the alleged "workers' state." And again the government backed down, revoking the price increases. At the same time the government imposed severe penalties on those it held mainly responsible for the strikes and riots.

Shortly thereafter a group of intellectuals of varying ideological persuasions ranging from Communist to Marxist formed the Committee for the Defense of Workers (known, from its Polish initials, as KOR), which beyond the task indicated by its name was to become during the next four years the focal and articulate mouthpiece for political dissent. For all its intermittent chicaneries against the dissenters, the Gierek regime refrained from wholesale and brutal repression of the kind that would have been resorted to between 1958 and 1970 (after Gomulka abandoned the liberal course he had taken during his first two years in power), not to mention the Stalinist era. There is no question that this unusual (by Communist standards) leniency was due, among other reasons, to detente. The now visibly malfunctioning Polish economy could be kept from total collapse only by increased borrowing from Western bankers. And it was largely by feigning political tolerance that the Gierek regime managed to convince the Western governments, notably that of the Federal Republic, that they should not discourage this alleged experiment in liberal Communism by clamping down on the flow of credits to Poland. In retrospect one stands amazed by the lack of foresight displayed by all the parties concerned. The Warsaw regime thought it could postpone the day of reckoning indefinitely, even when 80 to 90 percent of its hard currency earning was going to service its foreign debts. The Western governments acquiesced in their banks pouring money into a bankrupt economy. Statesmen otherwise as astute as France's Valéry Giscard d'Estaing and Germany's Helmut Schmidt cultivated Gierek, whom they believed to be very influential with the Kremlin, and for a Communist, quite pro-Western. Finally, the cardinal sin in the Communist catechism—"lack of vigilance"—must be laid squarely at the door of Brezhnev and Co. Their delight in the capitalists financing the Polish, and thus indirectly their own economy (the USSR was receiving, and at very favorable trade terms, Polish ships, locomotives, electronic equipment, etc.), blinded them to the clear portents of the coming crash and the potentially horrendous political consequences it might have, not only for Poland.

The relaxing—and, to a Communist hard-liner, demoralizing—atmosphere of detente had also its effects on several nonruling Communist parties. Those of France and Italy in particular had for years suffered from a sense of deep frustration. For all their mass membership and ability to win 20 to 30 percent of the vote in most parliamentary elections, they had for decades been barred from participating in the governments of their countries. Their ties with Moscow, and their formal commitment to "proletarian dictatorship" (read: one-party rule) made them suspect even in the eyes of many radical countrymen of theirs, and unacceptable as partners to other left-wing parties. The East-West truce, or at least its semblance, gave an added incentive to many Western European parties to try to attain greater respectability and convince their countrymen that Communists no longer stood for violent revolution or were mere puppets of Moscow, but on the contrary were ready to abide by constitutional and parliamentary rules and, while cherishing their historical ties with their Russian comrades, were entirely independent and solicitous of their own nations' interests in formulating their policies. This new transfiguration of Communism in the West became known as "Euro-Communism." The French Communist Party formally abrogated the harsher-sounding parts of its program, including the pledge to institute the dictatorship of the proletariat. This late-

in-the-day conversion to democratic procedures brought varying reactions from the non-Communists in the West. Many
were skeptical, remembering all the previous shifts and pledges
of independence by those parties, which in the end had turned
out to be just tactical manoeuvres designed to conceal or minimize their allegiance to and dependence on the USSR. To be
sure Western Communism was no longer simply servile to the
USSR, as in Stalin's time, when, as a French Socialist said
felicitously of his country's Communists, "They are neither on
the left nor on the right. They are in the East." Italian Communists, for example, have been quite outspoken in their criticisms of various aspects of the Kremlin's policy, its persecution of dissenters, the invasion of Czechoslovakia, etc. Yet,
when it came to the crunch, on most issues of international
politics the Western Communists somehow almost always
sided with the Soviets rather than with the Atlantic alliance
and their own governments.

Moscow's interest in Western Communists had always reflected their actual and potential utility to the Soviets. Since
the 1950s, at least, the Kremlin had given up on the possibility of either the Italian or the French party coming to power
on its own. There was a momentary flurry of interest in the
Portuguese Communists, who, after the collapse of the right-
wing regime, seemed to be in a strong position to stage a coup.
But in general the "fraternal parties" in the democratic countries were thought of as assets (to be sure) of the USSR, but of
limited value. Euro-Communism, therefore, was received by
Moscow with mixed feelings. Some in the Soviet hierarchy
shared the Western skeptics' view that it was but a tactical
move designed to induce other parties of the left to form alliances with the Communists and thus enable the latter to enter
the governments of their countries. And the Soviet Union could
not lose if Communists, even of the "Euro" variety, got a foothold in the French and/or Italian regimes. There was, on the
other hand, the melancholy possibility that even if they ini-
tially mouthed democratic slogans just for propaganda purposes, some Western comrades might imperceptibly come to
believe in them. Several foreign Communist leaders had al-

ready incurred the Kremlin's displeasure by their criticisms of the more repressive features of Soviet life and policies. Outstanding among these was the Italian party chief, Enrico Berlinguer. His pleas for a more liberal kind of Communism and his often-repeated assertions that his own party was committed to political pluralism had to lead to strong suspicions that Berlinguer actually meant what he was saying. Even more irritating was the Spanish leader, Santiago Carillo, whose explicit attacks on the residue of Stalinism in the Soviet system had to be censored from his speeches as reported in the Soviet press. Another source of irritation was the Western Communists' refusal to condemn emphatically and to break off interparty relations with the Chinese. All in all, by 1976 the Kremlin had decided that Euro-Communism should be discouraged and reprimanded as something which, though at present mainly a nuisance, might one day become a danger to the Soviet leadership of the "camp of socialism" and the world Communist movement.

The label "Euro-Communist" appeared to fit least well the French Party, many of whose leaders and activists gave repeated proofs of their nostalgia for those simpler (for a foreign Communist at least) times under Stalin. The French Communists' sudden abjuration of the class war and violence and their espousal of democratic values had thus to be viewed by their countrymen with more than a grain of salt. Even in the case of the Spanish and Italian parties, whose umbilical cord to the "Fatherland of Socialism" was stretched more precariously thin, one could ask, with Alexander Solzhenitsyn, why, if they were so critical of Moscow and condemned repeatedly its repressive actions at home and abroad, they still continued to maintain close links with it, attend the Soviet Party's conferences, join in its attacks upon American "imperialism," etc. This behavior was all the more striking because the Kremlin could deal with the foreign comrades quite unceremoniously whenever this suited its purpose. The French Communists had been trying for some time to mount a common front with the Socialists to oppose the government of President Giscard d'Estaing. But Giscard's continuation of de Gaulle's policy of putting a distance between France and the United States was found agreeable by the Sovet Union, which made no secret of its preference for him over any left-wing political formation. This brought expressions of anguish even from the most loyal French followers of the Soviet leadership. How could the Russians talk about proletarian internationalism when they praised Giscard's foreign policy, while it was really the French Communists who had been fighting for their country's independence from America? complained publicly and bitterly Jean Kanapa, known as one of the most pro-Soviet members of the French Politburo.[9] The Party's general secretary, Gaston Marchais, refused to attend its Soviet counterpart's Twenty-fifth Congress, and in fact did not visit the Soviet Union for more than three years thereafter, an unusually long absence from the birthplace of Communism for one of its foreign luminaries. But this was as yet in the nature of a family quarrel: the French might sulk, but on all issues that mattered they exhibited filial devotion to their Soviet elders.

Still, all things considered, Euro-Communism was another symptom of the fragility of the ideological rationale of Communism and hence that of the Soviet system. When out of power the foreign Communists could only sulk over Moscow's actions. But when and if masters of their own countries, wouldn't a Marchais or Berlinguer try to emulate Tito, or even Mao? It was easy to boast that capitalism had no future. But what future was there for Communism, when each of its fresh successes abroad brought new problems and potential dangers for the Soviet Union? In Eastern Europe it was not any ideological solidarity but the armed might of the USSR that secured the loyalty and the Communist character of the local regimes, and even there Albania had slipped the leash and Romania had managed to secure a degree of autonomy in its foreign policy.

Granted the premises of the Soviet Politburo, the overall world situation warranted two conclusions. First, the most logical and promising field for Russian expansion was in the Third World. Here the memories of Western colonialism

9. *Keesing's Contemporary Archives for 1976* (London, 1977), p. 27694.

handicapped efforts by America and its allies to prevent the USSR from moving in, and the fragility of the native regimes made it difficult for them to shake off the Soviets' influence once they accepted their aid.

Second, the complexities of the international situation made even more urgent than heretofore a continued and expanded military build-up by the USSR. In the Kremlin's view all the grueling political dilemmas just described could be alleviated, if not resolved, by growing Soviet might. For all their anti-Russian rhetoric, the Chinese had to be mindful of the considerable Soviet forces stationed along their border. Beijing would think twice before indulging in any provocation against Mongolia, Vietnam, or India. Further, investment in military hardware could bring handsome economic as well as political dividends. It was not purely commercial considerations that made the Europeans, and especially the West Germans, eager to expand trade and to lend money to the Eastern bloc. This was more and more viewed also as a means of propitiating the Soviet colossus, another insurance against a new Berlin crisis or worse. The silent persuasiveness of the growing numbers and might of Soviet ICBMs and tanks was also bound to discourage the West from too active a role in helping China industrialize and acquire modern arms. The Soviet navy, now first in the world in number of ships, was a vital adjunct in the expansionist policies in Africa and elsewhere in the Third World.

This extraordinary and many-sided military build-up was taking place against the background of decreasing Soviet economic growth. From an average annual rate of growth of 5.3 percent in 1965–70, it declined to 3.8 in 1970–75, and was to fall to 2.8 percent during the next five years.[10] Yet the amount of defense spending steadily increased in both absolute and relative terms. The peculiarities of Soviet accounting make it virtually impossible to pinpoint the actual amount of the budget spent on military and related purposes. Presumably President Carter was well informed when he stated in a speech

10. Quoted in Boris Rumer and Stephen Sternheimer, "Soviet Economy: Going to Siberia," *Harvard Business Review*, January–February, 1982.

in 1978 that "discounting inflation, since 1960 Soviet military spending has actually doubled . . . while our own military budget is actually lower now than it was in 1960."[11] Western experts have estimated that the Soviet military budget has been growing since the mid-1960s, at the rate of 3–4 percent per year. Yet in the face of this overwhelming consensus of opinion, Brezhnev at the Twenty-fifth Congress could say with a straight face: "The USSR has not increased its military budget; [on the contrary] it has steadily increased its appropriations for social and welfare purposes."[12] In the literal sense he was not telling a lie, for the budget appropriations for the *Ministry of Defense* have for some time hovered around the figure of 17 billion rubles, which according to the official (and largely meaningless) rate of exchange translates into a very modest $25 billion. But needless to say, no deputy of the Supreme Soviet would dream of asking how much was actually being spent on defense. And it is notorious that much of the Soviet military expenditure is included in the budgets of other ministries (e.g., research on and development of nuclear weapons has been traditionally carried on under the aegis of the inno-cent sounding Ministry of Medium Machines' Building).

The effort to conceal the real dimensions of the arms build-up and the burden it placed on the nation's economy was mo-tivated by political considerations both domestic and foreign. The Soviet citizen's appetite for more and better-quality con-sumer goods was obviously on the rise, and he would not have been enchanted by the true picture of how much was being spent on arms and in pursuit of the Kremlin's designs in Af-rica and other distant places. True, the average Russian has seldom balked at any sacrifices, no matter how great, when the security of his country is at stake, and the passage of more than thirty years has not dulled the memories of World War II and the terrible price the nation paid then for its government's lack of preparation against aggression. At the same time few Soviet men and women of the 1970s would have applauded

11. Quoted in *New York Times,* March 18, 1978.
12. *Twenty-fifth Congress of the Communist Party of the Soviet Union* [stenographic report] (Moscow, 1976), p. 46.

had they realized the price the country was paying for its rulers' ambitions and adventures in the Near East, Angola, and Southeast Asia, as well as the full cost of supporting Cuba's faltering economy and other burdens and dangers inherent in the regime's expansionist policies. There had been some food riots in the early 1960s, and while it was still inconceivable that the USSR might experience something similar to what had happened in Poland in 1970, the Kremlin was not taking any chances.

The regime was also eager to conceal from the outside world the strain its arms build-up was placing upon its economy. There was an inherent inconsistency in the Kremlin's desire to impress the West with its armed might while at the same time abjuring any aggressive and expansionist intentions. The key to resolving this dilemma, the Soviet leaders believed, lay in the continuation of the SALT process. As long as the USSR and America had an agreement on strategic weapons, or even continued negotiating about one, the West was unlikely to feel the need to match or surpass the Soviets' military effort. Masters of the art of having one's cake and eating it too, the Soviets wanted the US and its allies to be indeed somewhat frightened of their growing armed might, but not to the point where those countries would embark on a vigorous arms build-up of their own. The US government's economic analysts were thus faced by the difficult task—a veritable jigsaw puzzle—of putting together odds and ends of incidental information into a realistic assessment of Soviet defense spending and then translating it into dollar terms, the result being a figure five or six times bigger than the "official" defense budget.

The quadrennial political earthquake otherwise known as presidential elections preempted the US scene in 1976. This rather strange (in their view) American rite has always been a cause of perplexity to the Soviet rulers. Many items on the agenda between the two countries had to be held in abeyance since Brezhnev's American counterpart had to be preoccupied with the feelings and aspirations of the denizens of New Hampshire's hamlets and New York's inner city, leaving him little time and less energy for the intricacies of SALT II or of balancing US policies between the USSR and China. Another inconvenience inherent in the elections was the possibility that having just come to know and learn to gauge the reactions of those who guided US policies, the Kremlin might suddenly find itself faced with the necessity of dealing with an entirely new team in Washington. One could never exclude the possibility of an unpleasant surprise—American politicians being so fickle and public opinion in that strange country so often unpredictable in its reactions to foreign affairs. When the final contestants ultimately emerged from those disorderly primaries and conventions, the Kremlin appeared to have no reason to worry. President Ford had survived an assault by the rightwing element in his party, one of whose main complaints had been America's weakened defense posture and overall international position vis-à-vis the USSR. The Democrats' champion ran largely on the residue of anti-Vietnam and post-Wategate emotions, and he proposed to reduce US defense expenditures. Both candidates eschewed a militant anti-Communist position.

But long and bitter experience had taught Moscow that what the American politicians said in the heat of a campaign was not a reliable guide to what they actually would do once in power. The Republicans tended to talk more harshly about the Soviet Union, but once in power they tended, as behooved true capitalists, to be more cautious and amenable to compromise. The Democrats were on the whole a motley and excitable crowd, and hence, in the opinion of the sedate and deeply conservative men of the Kremlin, quite unreliable and potentially dangerous to deal with when in power.

The election of Jimmy Carter must then have been greeted by Moscow with mixed feelings. For his part he entered the White House, as had practically all American presidents since World War II, with high hopes of effecting a major and salutary breakthrough in US relations with the Soviet Union. The new administration was staffed, especially at its secondary levels (assistant secretaries of state, etc.), by people who still shared the traditional American belief (others might call it an illusion) that one of the main keys to improved relations with the Communist world lay in following resolutely progressive policies and displaying sympathetic understanding toward left-of-center regimes and movements in the Third World. In 1977 many Americans still felt that the Soviets should react positively to the facts that US policies were being run by those who had been critical of our past involvements with foreign reactionaries, that the new president had pledged to cut defense spending and to pull US troops out of South Korea, and that Washington now abjured power politics as allegedly practiced by the late administration's secretary of state.

Typical of what the Soviets termed "the new relationship of forces in the world" was the nervousness expressed in some American circles that the Soviets might "test" the new President by contriving an international crisis. But as suggested above, it was the Russians who have usually tended to become nervous at each change of guard in Washington. Unlike Khrushchev, who was indeed not above trying to bully Western statesmen, especially when freshly in office (e.g., as with President Kennedy in Vienna), Brezhnev and his colleagues could be expected to be quite circumspect, at least until they had taken the measure of the new helmsman of US foreign policies. The new President, it was clear, would need some time to familiarize himself with the intricacies of world affairs. His main foreign adviser conformed to the usual pattern for American secretaries of state: a distinguished lawyer with extensive experience in governmental affairs. The presidential assistant for national security, a position which since the early 1960s had rivaled (and with Kissinger surpassed) in importance that of the secretary of state, was Polish by birth and a Kremlinologist by profession, neither characteristic particularly reassuring to a Soviet official.

The first year of Carter-Brezhnev coexistence dispelled much of the new administration's original optimism about being able to effect a major and beneficial change in US-Soviet relations, as well as in those with some other Communist states. The new Washington team's openness in managing foreign affairs (as against all those secretive ways attributed to Kissinger), its populistic image and rhetoric seemed to have little effect on the Communist rulers. And for its own part the Kremlin, much as it must have been reassured by the President's obvious eagerness to patch up detente, could not but become both apprehensive and irritated by the new American emphasis on human rights.

It is very likely that Carter's previous utterances on this subject had been treated by the Soviets as typical "campaign oratory" designed to refute the charge by his opponents that once in office he would be "soft" on Communism. It must have appeared quite incongruous to the Soviet mind, and perhaps not entirely unjustifiably so, that the same man who had expressed the view that the Americans were unduly worried about the Soviet-Communist danger, who proposed to reduce the defense budget and hoped to open channels of communication with Havana and Hanoi, could be entirely serious when promising that his government would keep an unceasing vigil over what was happening to the dissidents in Russia and in other countries. Yet once in the White House Carter embarked on a number of actions and gestures which must have exasperated the Kremlin, as well as disturbed some of the more conservative professionals in the State Department. Academician Andrei Sakharov had long been an unremitting and severe critic of his country's regime, which because of Sakharov's world renown as well as his standing within the Soviet scientific community found itself unable to deal with him as it had with others of his ilk, i.e., by imprisoning or expelling him from the USSR. He now addressed a letter to the new President, asking him to persevere in his advocacy of human rights for his countrymen, a letter that was publicly acknowledged and answered by Carter. Something worse followed: largely through the good offices of the US, a swap was arranged of Louis Corvalan, a Chilean Communist, and Vladimir Bukovsky, a very articulate Soviet dissenter, both hitherto in their respective countries' jails. After Alexander Solzhenitsyn's expulsion from the USSR some years before, President Ford had been careful to eschew any personal contact with the great writer. Carter met with Bukovsky in the White House.

The Soviet press and officialdom put on a great show of

outrage. The Soviet ambassador, Dobrynin, lodged a formal protest with the State Department, stating that his government "rejects resolutely any attempts to interfere in its internal affairs, based as they are on the fictitious pretext of [protecting] human rights." [13] The Soviets, the official Moscow announcement said, could themselves say quite a few things as to how human rights were faring in the US, and expatiate on such subjects as "unemployment, which has afflicted millions, racism, discrimination against women, restrictions on civil liberties, the rise in crime and delinquency, etc." But the USSR would not dream of bringing up such subjects, not only out of delicacy, but because "to inject such issues into intra-governmental relations would only complicate the solution of those problems which can and ought to be subjects of negotiation and cooperation between the two countries." The fact that the "convicted criminal" Bukovsky had been entertained at the White House was felt to be a special affront to the national dignity of the USSR. On March 4 *Pravda* carried a selection of a number of foreign criticisms of this allegedly unprecedented step. It cited approvingly an American editorial that supposedly found Carter's gesture provocative and contrasted the new administration's alleged solicitude for human rights in the USSR with its attitude towards what was happening in Nicaragua, where "with the tacit approval of the CIA . . . bloody terror is being practiced, with the soldiers of the dictator Somoza, who is being supported by Washington, destroying whole villages and murdering people, including women, the elderly and children." There followed a veritable flood of articles and editorials pointing out how under the cover of "human rights" American correspondents in the USSR and CIA agents had been trying to enlist Soviet citizens as spies.

Genuine as was the Soviet leaders' indignation, it is unlikely that it reached the heights suggested by the furious tone of the press campaign. Some of the more cynical among them may well have concluded that Washington's bold rhetoric about human rights was designed primarily to conceal its con-

13. *Pravda,* February 19, 1977.

fusion and uncertainty over how to deal with the Soviet challenge in the Third World, as well as to appease the mounting
Congressional concern about the Soviet arms build-up. Its
censoriousness about the Soviets' domestic practices notwithstanding, the administration was more than eager to renew
negotiations with Moscow about all the issues outstanding between the two countries, and especially to conclude a new
strategic arms pact, for SALT I was due to expire in October
1977. At its secondary level the new Washington team included quite a few people who had been active in the peace
movement of the 1960s and strongly critical of many aspects
of US foreign policy since World War II. Thus there were great
hopes that a new spirit could be injected into American relations with Third World regimes, including the Communist
ones, and that a constructive dialogue might be established
with Vietnam and Cuba, even though neither was particularly
known for its observance of human rights. With respect to
these, the administration, taken aback by the violence of the
Soviet reaction to its gestures vis-à-vis Russian dissidents,
hastened to explain that its solicitude in this respect was not
limited to any particular state or type of regime. "By announcing the sanctions against Argentina, Ethiopia and Uruguay,
Secretary of State Cyrus R. Vance underscored the President's
assertion yesterday that his concern about human rights was
a global policy not directed solely at the Soviet Union." [14]
Needless to say, such statements could hardly be expected to
mollify the Kremlin, though it must have watched with wry
amusement how Washington's preaching of universal virtue
was creating new strains in its relations with a number of non-
Communist states.

The jarring dissonance about human rights, as well as the
perhaps inevitable confusion that follows a change of administration, did not deter Carter and his advisers from trying to
open substantive negotiations with the USSR very shortly after
his inauguration. In March Secretary Vance was dispatched to
Moscow bearing a new set of proposals which, it was hoped,
would provide a framework for SALT II. In retrospect it is clear

14. *New York Times*, February 25, 1977.

that this haste was unfortunate. Considerations of protocol, always an important point with the Russians, called for Gromyko to initiate high-level contacts between the two powers, by coming to Washington and paying his respects to the new President. The Americans' display of overeagerness for an agreement provided another incentive—quite apart from its irritation over the dissident issue—for the Kremlin to prove dilatory or worse. Finally, much as they do it to other people, the Soviets dislike having surprises sprung upon them.

To be sure none of Vance's proposals was entirely new, or contained anything that could not have been anticipated by the Kremlin. But evidently no effort had been made to probe the Soviets' reaction before the secretary of state was dispatched on his mission. One alternative plan he presented to his Russian hosts was that SALT II negotiations should adhere to the guidelines laid down in Vladivostok more than two years before. Those guidelines provided for a ceiling of 2400 nuclear missiles and heavy bombers for each power, with not more than 1320 to be MIRVed. Two categories of weapons not included in the proposed quotas of the Vladivostok aide-memoir, as it was officially styled, were cruise missiles and Backfire bombers, each capable of carrying nuclear arms. Both countries were developing the former, but the US was known to be well ahead in cruise missile technology. "Backfire" was the Western code name for the Soviet bomber that US experts claimed had an intercontinental range and hence should be included among strategic weapons; both propositions were stoutly rejected by the USSR. It was therefore agreed to postpone the discussion of cruise missiles and the Backfire until further talks.

Secretary Vance and his fellow American delegates were thus both astounded and chagrined when their Soviet interlocutors declared categorically that the Vladivostok conferees had agreed to include the cruise missile question in the proposed nuclear arms agreement. As Gromyko was to say subsequently in a press conference, "They attempted to argue that cruise missiles were not included in the Vladivostok agreement, that those missiles, if you please, should be subject to no limitations and that the Vladivostok plan referred only to ballistic missiles. We resolutely rejected such arguments. The question was addressed at Vladivostok; there was no green light given there to [unlimited] development of cruise missiles." [15] As for the Backfire bomber the Soviets declared that it was ridiculous to consider it a strategic weapon; everyone knew it did not have the range to reach the US and to come back to its home bases! (Which of course overlooked the possibility of Backfire being refueled in the air or, say, in Cuba.) In other words, there could be no question of *not* including cruise missiles in SALT II negotiations, while the Backfire was completely irrelevant to them!

If the American first option for SALT II was thus brusquely rejected, then Secretary Vance's alternative plan produced a veritable, though mercifully not nuclear, explosion on the Soviet side. Again to quote Gromyko, "They now propose to reduce the strategic-weapon carriers to 2000 or 1800, and those MIRVed to 1100–1200. Furthermore they want us to dismantle half of those missiles of ours that some people in the United States happen to dislike. [That type of missile] they describe in different ways, sometimes 'too heavy,' sometimes 'too effective.' The Americans don't like it, and therefore the Soviet Union ought to dismantle half of those missiles. One must ask: can such a one-sided approach to the issue lead to an agreement?" [16]

Indeed no one knowing Brezhnev, Gromyko, and their colleagues could have expected them to welcome Washington's alternative approach to SALT II. As indicated above it envisaged a considerable reduction in the nuclear launcher ceilings projected in Vladivostok. And once again nuclear arms decisions made by the Pentagon way back in the 1960s came to haunt the American side, but this time with much greater poignancy than in 1972. As we have seen, the rationale then for not trying to match the Soviets' biggest missiles was that the American ones were more accurate and that this country's technology for equipping them with MIRVs was supposedly

15. Quoted in *Pravda*, April 1, 1977.
16. *Ibid*.

much superior. But in 1977 the USSR disposed of upwards of 300 giant SS-18 missiles, each capable of being equipped with up to ten independently targeted warheads. Should the USSR ever resort to a preemptive strike, those giants by themselves, claimed some American experts, could annihilate most of the US land-based ICBM force. Hence the American proposal that in the interest of equalizing deterrence the Soviet Union should reduce the number of its superheavy missiles by half, which would still leave it with about 150 of those monsters, against none for the US.

With the Soviet leadership refusing even to discuss either of the US alternative plans, there was little for Vance and his colleagues to do but leave Moscow in some discomfiture. Yet for all the frostiness, if indeed not outright rudeness, exhibited by the Soviet side during the talks—they could hardly be called negotiations—the Soviets were careful to leave the door ajar for an eventual agreement. "It was stipulated that these problems would continue to be discussed by both sides, with the participation of technical experts." [17] In Washington Carter had to acknowledge the failure of his first major diplomatic initiative. No, he did not think, he said in a press conference, that this had anything to do with the Soviets' vexation over his human rights stand. And he repeated that both ex-President Ford and Kissinger had confirmed that the US had never agreed at Vladivostok to forego the deployment of cruise missiles. [18]

There can be little doubt that the Soviets had chosen deliberately to teach the new administration a lesson and to throw it off balance insofar as the future SALT negotiations were concerned. How else could one interpret Gromyko's biting sarcasm at his press conference after Vance's departure? "In fact what would happen if each time you get a new leadership in some country it should proceed to cancel all the positive achievements previously reached with other states? What kind of stable relations could you then have in [international] relations?" And in answer to a correspondent's question as to whether the issue of human rights did not affect adversely the

17. From an official communique, *Pravda*, April 2, 1977.
18. *New York Times*, March 31, 1977.

Moscow talks, the Soviet foreign minister did not choose to be evasive or diplomatic. "All that talk which has lately been let loose in the United States about 'man's [sic!] rights' . . . obviously does not help, it poisons the political climate. And how could that help in the resolution of other problems, including those touching on strategic weapons? No, this does not help. It does harm. . . . I might also say that we do not need any teachers [from abroad] when it comes to the internal affairs of our country." [19]

For all his real or feigned indignation, Gromyko was careful to state that the negotiations would proceed and should eventually lead to an agreement. But the Soviet bargaining position now hardened. The new SALT could not be limited to just setting quotas on US and Soviet strategic weapons. Two other issues would have to be considered. "First . . . the question of transfers of strategic weapons to third countries . . . nothing should be done to evade this issue and the signing of an agreement concerning it. We already have a concrete proposal concerning this matter." For all the intentional obscurity of this phrasing, no one present could miss what Gromyko had in mind. SALT II must make it clear that the US was not allowed to provide any other country (read: China) with nuclear technology or weapons.

And while nobody could doubt Soviet earnestness on the last point, the other fresh item injected by Gromyko into the SALT agenda was obviously introduced for bargaining purposes. At Vladivostok the Soviets had refrained from raising the issue of America's *tactical* nuclear weapons, i.e., those shorter-range missiles and bombers placed in Europe under NATO. But now that the Americans had tried to introduce new elements into the negotiations, the Soviets felt entitled to alter their position too. "We have the right to raise the issue of dismantling the American forward bases. This concerns nuclear submarines, bombers equipped with atomic weapons, and aircraft carriers, all of them operating in or near a certain region of Europe. (And you all know about what region we are speaking.) Call it what you will; the hardening of our position or

19. *Pravda*, April 1, 1977.

that we have changed our position. But I repeat, this question now arises because of the Americans' latest proposals . . . this touches on our security and that of our allies." [20] Here Gromyko's meaning was quite clear: if you Americans pester us about our SS-18 and the Backfire, we will demand that you dismantle your tactical nuclear weapons in Central Europe and even remove such weapons from your aircraft carriers and submarines in the North Sea and the Mediterranean.

To use the vernacular, Washington had led with its chin and the Kremlin did not fail to deliver the blow. From now on the Carter Administration would stagger through the SALT II negotiations, inhibited from pressing the Soviets to make substantive cuts in their forces—for they in turn would demand virtual nuclear disarmament of NATO—while at the same time subjected to a mounting outcry by the "hard-liners" at home that any agreement that would preserve and in fact enhance Moscow's superiority in megatonnage could lead to a disaster for the West.

The initial American misstep was bound to affect—from this country's point of view, unfavorably—other aspects of the relations between the US and the USSR for the balance of Carter's term. The brazen tone of Gromyko's remarks reflected the Kremlin's conviction that the administration, having committed itself to a speedy conclusion of SALT II, as well as to a general improvement of US relations with Moscow, could not now abandon this course precipitately without incurring considerable political risks. For a large portion of the American electorate, the only alternative to detente was "cold war," a term few could now define with any precision, but which conjured up the prospects of vastly increased defense expenditures, renewal of the draft, fresh Vietnams and Koreas. America's allies in Europe were also eager for the Soviet-US dialogue to continue. They would deplore any drastic turnabout in the two superpowers' relations. With their economies also beginning to feel the effects of recession, the Western European states could not be expected to welcome any action by Washington that might constrain them to increase their own de-

20. *Ibid.*

fense spending and which would also undercut their still-expanding trade with the Eastern bloc. West Germany, increasingly the pivotal state in the Atlantic alliance, was especially apprehensive about the possibility of any sharp deterioration in East-West relations. Bonn had come to prize highly the increased contacts detente had allowed between the two Germanys, as well as the fact that it was now easier for ethnic Germans still in Poland and the USSR to return to their homeland. Many European leaders, while deploring (especially in private) what they saw as the American policy-makers' naiveté about the Russians, still did not want them to become realistic to the point where that might trigger another crisis over Berlin or something worse.

There were thus numerous factors both domestic and intra-alliance that urged Carter and his advisers to persevere in their patient, guardedly optimistic attitude vis-à-vis the USSR. Some in Washington still believed in the magic of "summitry" and thought that many of the recent difficulties and misunderstandings would be dispelled when, as seemed eventually inevitable, the President and Brezhnev met face to face. Others, on the contrary, placed great hopes on something they thought could not be long delayed: a wholesale change of guard in the Kremlin. Brezhnev was over seventy, and evidently not in very good health, and few of his Politburo colleagues were much younger. Once the aged oligarchs trained in Stalin's school were gone, their successors would likely be men with a different set of priorities: more solicitous of the needs of their own people and less prone to cause mischief abroad. And so for the moment one had to continue trying to meet the Russians halfway on SALT II, and be philosophical about their ventures in Africa. There the Soviets were bound to discover sooner or later what the Americans had already learned in Vietnam: even for a superpower it was futile in this day and age to try to impose its will on nations thousands of miles away from its own borders. Soviet forays into the Third World could be best countered not through the CIA or a return to Cold War rhetoric and posturings, but by enhanced American understanding and help to the new nations. The North-South dialogue, i.e., increased aid and sensitivity to the needs of the underdeveloped countries on the part of the industrialized ones, was, according to the proponents of this line of thinking in Washington circles, desirable not only for its own sake, but also because it would lessen East-West tensions.

There thus seemed to be no compelling reason for the Kremlin to become more restrained in its foreign policies. The USSR held doggedly to the main lines of its announced position on SALT II. Any attempt by the American negotiators to reopen discussion of one of the modifications contained in Vance's March 1977 proposals was stoutly resisted by the Soviet side and denounced as a US design to extract one-sided concessions. The Americans, Moscow insistently held, had to reconcile themselves to the fact that the days of their nuclear superiority were gone forever. The Soviets' own proposals, they argued—i.e., essentially the Vladivostok plan with some alterations—were eminently fair, for they provided for essential parity and mutual assured deterrence. The USSR neither possessed nor sought the capability to launch and win a nuclear war. It was the lurid imagination of some Americans that made them draw up those fantastic scenarios of a Soviet preemptive strike destroying most of the US land-based ICBMs and thus crippling this country's power to retaliate. And if such nonsensical and slanderous fabrications scared many people in the West, it was certainly not Moscow's fault.

While staunchly defending what they presented as the status quo where nuclear arms balance was concerned, the Soviets were strongly critical of the US efforts to assert this principle in other spheres of international relations. The world was changing, and it was both childish and futile for the West to ascribe those changes of which it did not approve to some sinister Soviet machinations and expansionist designs. Take Africa, whose nations were determined to eliminate the last relics of colonialism and to combat neo-imperialism. It was natural that in their struggle for full political and economic emancipation they looked to the socialist camp for sympathy and support, and equally natural that if thus solicited by the new nations, neither the Soviet Union nor its Communist friends could begrudge whatever help they could give. The evanescence of Western influence in Africa and other Third World areas was thus part of the inevitable historical process and could in no way be blamed on some plots allegedly hatched in Moscow. And by what right could Washington or any other Western capital object to the Soviet Union establishing close links with such independent countries as Angola, Somalia, and South Yemen? Didn't France, for example, though it had officially dismantled its empire, still exercise a form of tutelage and maintain its troops in several of its former colonies? And it was notorious that Colonel Mobutu's dictatorial government in Zaire managed to stay in power very largely because of America's support. And the British! For all their alleged withdrawal from "east of Suez," Great Britain still helped such reactionary regimes as those in Oman and the United Arab Emirates to combat their countries' progressive and patriotic forces.

At the level of rhetoric the Carter Administration found it difficult to cope with the Soviet arguments. Practically from the beginning of the republic, American statesmen have deplored and opposed the notion of spheres of influence as essentially immoral. (And this country did not quite realize it had one of its own until it was definitely breached with the establishment of a Communist regime in Cuba.) It would thus have been awkward to argue that the Russians had no business moving into Africa because, well, they had not been there before. At the practical level, countering Soviet activities on that continent became more difficult following the recent legislative curbs on covert activities by the CIA and others. The issue was further muddied by the Soviet use of surrogates in advancing their political goals in those distant lands: Cuban troops, East German technicians, etc. The overall picture was thus quite confusing, not only to an average American, but sometimes even to a high official. It was the US chief representative in the United Nations who opined on one occasion that the Cubans were performing a useful, stabilizing function in Angola. And the latter's government, while drawing ever closer to Moscow, signing a formal treaty of friendship and mutual assistance with the USSR, and soon becoming an associate member of Comecon (the Communist analogue of the Common Market), was still circumspect enough to allow an American oil concern to operate on its territory and to deny allegations of any undue dependence on the Russians.

The Soviets should have known that the apparent ease with which they seemed to be scoring points against the West could both be deceptive and have dangerous consequences. Experience taught that a moment might come when the sum of American discomfitures and apprehensions reached critical mass, triggering one of those sudden swings of the public opinion pendulum that have periodically occurred in this country's history, leading to a sharp anti-Soviet reaction. But in 1977 the possibility of such a drastic shift appeared, at least in the Kremlin's view, quite remote. Consequently it embarked in 1977–79 on a number of ventures that departed from the rather cautious pattern of Soviet foreign policy during the early period of Brezhnev's regime.

To be sure Soviet penetration of the Third World, especially Africa, had been gathering momentum, and it would have required an unusual degree of self-restraint on the part of the Kremlin to arrest it. The turbulent politics of the area on the one hand, and the West's disarray and indecision on the other, offered alluring temptations to extend Soviet influence and to undercut that of the West.

In the Horn of Africa Moscow had for some time maintained close relations with the regime of Somalia. Though sparsely populated and unfavored by nature, this country because of its location had achieved considerable strategic importance in post-colonial Africa. Like its neighbor across the Gulf of Aden, South Yemen, Somalia became a recipient of Soviet economic and military aid, and in return provided the USSR with naval and airfield facilities in an area close to some of the West's main sources of oil. Somalia's much larger neighbor, Ethiopia, had on the contrary traditionally depended on American help, both in the military sphere and towards whatever efforts at modernization this essentially feudal monarchy saw fit to undertake. The Ethiopian revolution of 1974 ushered in an era

of violent internal changes that soon had international repercussions. The overthrow of the monarchy was followed by a
number of bloody coups. The latest, in February 1977, had
brought to power Col. Mengistu Haile Meriam, who both
through disposing of his rivals and by assuming the presidency of the ruling military junta, known as Derg, became the
country's virtual dictator. Coups and subsequent authoritarian
rule by (usually young) military figures had been quite frequent in Africa's recent history. What made the Ethiopian case
somewhat unusual was the fact that the Mengistu regime proclaimed its adherence to socialism, and not of the "African"
variety as had hitherto been customary, but of the Marxist-
Leninist one.

The first stages of the Ethiopian upheaval were watched
warily by Moscow. The old Emperor, for all his links with the
US, was not unfriendly toward the Soviets, and as Africa's
senior statesman had been vigorously courted by them, visiting the USSR on three occasions. But the main element in that
initial caution must have been uncertainty as to whether the
new regime, indeed the state as a whole, would long survive.
The sanguinary practices of Derg had led to outbreaks of rebellion in several parts of the country. Despite their consideration for Haile Selassie, the Soviets had for years discreetly
supported a secessionist movement in Eritrea, which had
erupted when the Emperor reneged on a promise of autonomy
to that former Italian colony, which the Allies following World
War II had unwisely turned over to Ethiopia. And Moscow's
protégé, Somalia, had never concealed its claims to Ogaden,
part of Ethiopia but inhabited mostly by Somalis.

As in the initial stage of the Soviet romance with Castro's
Cuba, so in the case of Ethiopia: until 1977 Moscow was in
the position of the wooed, rather than that of the wooer. While
quite skillful at physically liquidating his own and Derg's opponents, Mengistu proved unable to stop the momentum of
his country's territorial and economic disintegration. The government's efforts to apply "scientific socialism" in a hurry had
led to famine in several areas. The government's armies, badly
trained and demoralized, were being trounced not only by

rebels in Ogaden, now assisted by regular Somali troops, but also by the Eritrean insurgents, as well as a variety of guerrilla forces. It was clear that only an outside power could rescue the bloodstained regime and prevent Ethiopia from falling apart.

A trickle of American economic help had continued to reach the country until the end of 1976. But quite apart from any ideological scruples they might have had, Mengistu and his entourage must have realized quite early in the game that it was hopeless to expect the United States to save them from their numerous predicaments. No American administration, even one much less committed to the defense of human rights, could or would have been allowed by public opinion to provide massive help to a regime that was notorious for practicing indiscriminate terror against its own people, and for settling its internal squabbles in a manner reminiscent of Murder, Inc.

The Soviets on the other hand were ready by the spring of 1977 to embrace and bail out the Mengistu regime. The colonel visited Moscow in 1977 and was received by the highest Soviet officials. His visit initiated a massive flow of Soviet arms to Ethiopia which, it was estimated, amounted during the next twelve months to more than $1 billion.[21] By the summer Soviet and Cuban military advisers were already instructing Derg's troops. Then, as the Ethiopian army still continued to suffer reverses, especially in Ogaden and Eritrea, the Kremlin repeated its Cuban gambit. In February 1978 the national security adviser, Brzezinski, asserted that between ten and eleven thousand regular Cuban troops were fighting in Ogaden. Overall command of the military operations against the insurgents and Somalia was in the hands of senior Soviet officers, the top one being identified as Gen. Vasili Petrov, previously deputy commander-in-chief of all Soviet ground forces.[22]

The initial Soviet reluctance to sponsor the Addis Ababa regime was probably due not to any squeamishness over its

21. Colin Legum and Bill Lee, *The Horn of Africa in Continuing Crisis* (New York, 1979), p. 13.

22. *Ibid.*, pp. 14–51.

internal policies, but to the realization that such support would carry certain risks and costs. For one, it was bound to jeopardize the Soviet position in Somalia. Rather awkwardly, the Soviets tried to get over this difficulty by suggesting that the two warring countries should resolve their territorial dispute by forming a federation. But this ingenious suggestion proved impractical. President Ziyad Barre of Somalia made a hurried trip to Moscow, but his pleas that old friends should not be jettisoned in favor of new ones received short shrift from the Kremlin. Ethiopia was simply a much bigger prize; both its territory and population were much larger than Somalia's and its potential strategic and economic importance dwarfed that of the largely desert land of the Somalis. The latter's original success in the fighting in Ogaden was largely due to the military instruction and equipment that its army had been receiving from the Soviets since 1967.

But the main Soviet concern had to touch not on what Somalia, with its population of three million, might do, but on the possibility of wider international repercussions of yet another massive injection of Soviet power into an intra-African conflict. Most Arab states, including some with pro-Moscow regimes, supported Somalia's and the Eritrean rebels' claims against Ethiopia, and viewed with apprehension the Soviets' games in Africa. And on one occasion in 1977, what was in all likelihood a Soviet probe in another region of the continent had been in fact defeated through a joint Arab-West effort. In April, armed opponents of Colonel Mobutu's regime who had found shelter in Angola invaded a province of Zaire. In answer to Mobutu's appeals for help, the French provided planes that airlifted Moroccan soldiers who assisted Zaire's troops (allegedly including Pygmies armed with bows and arrows!) in squashing the rebellion.

But whatever hesitations the Soviets may have had about a full-scale commitment to its Ethiopian friends, they must have disappeared following the regular Somali army crossing into Ogaden in July. This undoubted violation of an international border put Washington into a quandary. It had previously hinted at its readiness to supplant the Soviets as a friend and supplier of military aid to Somalia, but now felt it could not openly help what was from the legal point of view the aggressor state. The US role during the latter stages of the imbroglio was confined to appealing to both sides to resume the straight and narrow path of international virtue: to Somalia to respect Ethiopia's territorial integrity and desist from invading, to the Soviets to pull their own military people and the Cubans out of Ethiopia.

Hoping to soften the West's legalistic posture, Barre in November 1977 denounced the Soviet-Somali Treaty of Friendship and Cooperation of 1974, expelled the remaining Russian advisers, and terminated the Soviets' use of the port of Berbera. This last reprisal was probably the most painful one for Moscow, since the Soviet navy had turned the base into a mainstay of its presence east of Suez, equipping it with oil and weapon storage facilities, airfields, etc.

But the Somalis' actions only increased the Soviets' determination to teach them a lesson, without at the same time softening the resolve of the United States not to get involved in the conflict. An official White House statement of November 14 reiterated the decision not to supply arms to either of the warring sides. Somalia's action in ousting Soviet advisers was indeed praised by Washington as placing Somalia "in a far better position to pursue a truly nonaligned foreign policy."[23] But this praise could be of little comfort to President Barre, whose "nonaligned foreign policy" now resulted in a series of defeats for his army at the hands of the Cuban-Ethiopian forces. By February the shoe was definitely on the other foot: it was Somalia that faced the danger of being invaded and quite possibly overrun—which indeed might have led to an Ethiopian-Somali "federation," but hardly of the kind some Soviet publicists had envisaged before the start of the conflict. But the supply of Cuban soldiers was not inexhaustible, and they were also needed elsewhere: on other Ethiopian fronts, in Angola, and in South Yemen, whose political leadership, though friendly to the USSR, was evidently thought not to be

23. Quoted in New York Times, November 15, 1977.

friendly enough, and was in July 1978 overthrown and replaced by a more trustworthy one.

But the main reason that Moscow eventually agreed not to
carry the war into Somali territory lay in the growing rumblings in Washington over the Cubans' activities in Africa, and
in the fact that because Somalia was a member of the Arab
League, its absorption at this particular moment would have
strained the pro-Soviet attitudes of regimes such as Syria's and
Iraq's. Once Somali territory had been invaded, Washington's
legalistic scruples presumably would have lessened. And even
more important, there were people in the Carter Administration who talked about conditioning the continuation of SALT
negotiations upon the Soviets' restraint in Africa. In March
the last regular Somali units were pulled out of Ogaden, and
Moscow let the State Department know that though it did not
of course control the actions of a sovereign state like Ethiopia,
it had good grounds to believe that the Cuban-Ethiopian forces
would stop at the border. As for the general problem of the
Cuban forces in Africa, yes, the USSR hoped that they could
go home or be reduced, but of course this was something that
was up to the respective governments, i.e., Ethiopia's and Cuba's. Later on in the year Gromyko admonished once more the
Western critics of the Kremlin's actions in Africa: "Some people in Washington and other NATO capitals blame the Soviet
Union for everything that is not to their liking . . . apparently
they need to create some kind of screen to cover their actions,
to hide what happened in Zaire." [24] As against those nefarious
activities of the West, the USSR, its foreign minister continued, should be given great credit for its altruism and self-
restraint: "Every fair-minded person should have a good word
for the Soviet Union, which has assisted a victim of aggression with arms—I repeat, with arms. Not one Soviet soldier
with a rifle is in Ethiopia." Soviet officers, of course, carry
only side arms.

The varied mosaic of African politics, with its often bizarre
and rapidly changing political alignments and realignments,

24. Quoted in *Pravda*, June 3, 1978.

thus seemed to be made to order for the Soviet policy-makers' attention and exploitation. The use of Cuban troops in pursuance of the Kremlin's designs had by 1978 become such an accepted, if not applauded, fact of international life that it became an argument in pressing Rhodesia's (now Zimbabwe's) white minority government to hurry up and hand over power to the country's black majority. Otherwise, some *Western* politicians argued, the local guerrillas might soon be joined by the Cubans, and Moscow's long reach would then foreclose the possibility for any democratic compromise solution of the Rhodesian crisis.

While seeking through military means to acquire a new client, the USSR was at the same time employing diplomacy to minimize the consequences of losing an old one. Having in 1976 abrogated the Soviet-Egyptian friendship treaty, President Sadat drew even closer to the US. What was in some ways even worse from the Kremlin's point of view, the Egyptian leader gave indications of seeking an accommodation with Israel. The Arab-Israeli conflict had provided the opening for the Soviet Union to establish a foothold in the Near East. At other times Moscow might have welcomed a certain lessening of tension in the region; as we have already seen, actual war there would always raise the spectre of a US-USSR confrontation or, contrariwise, would demonstrate that Moscow was less than wholehearted in its support of the Arab cause. But any far-reaching accommodation between the two sides threatened the Soviets with the loss of most of their leverage in an area that had become crucial to the overall design of their foreign policy. The defection of Egypt was especially painful. In immediate terms it meant the loss of an ally that militarily and in terms of population was the most important state in the Middle East. But its potential implications were at least equally unpleasant. Moscow felt, and justifiably, that it had saved the Sadat regime from a military disaster in 1973. It had poured billions into what it assumed constituted a bastion of Soviet influence, not only in the area but insofar as the whole Third World was concerned. What lessons would other Third World leaders draw from the example of the Egyptian dictator who, having used the Russians, proceeded to discard them, and enter upon a sinful liaison with the United States?

The USSR thus began a series of intricate diplomatic manoeuvres designed to undermine Sadat's position as the spokesman for the Arab cause, as well as to make sure that any attempted approach to the solution of the Middle Eastern conflict would involve Soviet participation—something which would mean long and dilatory negotiations, in the course of which the Soviets would be able to teach the presumptuous Egyptian statesman some lessons.

The Middle Eastern question figured prominently in the discussions conducted by Gromyko with high American officials on the occasion of his visit to the US in September–October 1977. This time Gromyko's language and behavior were quite different from the sarcastic, almost insulting tone he had used during his American counterpart's visit to Moscow a few months before. Instead of making slighting allusions to the new administration, the foreign minister in his press conference at the UN stressed the positive results of his talks with President Carter and other high American officials. There had been progress, he said, on the strategic arms limitations talks. "Today in this respect the situation is better than it was yesterday." And while there was no chance of the two powers reaching a SALT agreement by the October deadline, they both pledged to continue to observe the limitations stipulated under SALT I until such time as a new treaty became possible.[25] Both sides had also discussed in a friendly and constructive manner a multitude of other international problems, including those of the Middle East and Africa. Concerning the latter, the Soviet minister skipped nimbly over the situation in the Horn of Africa, where, as he spoke, the Russians were airlifting Cuban troops and war materiel to their friends in Ethiopia. No, concerning Africa, "the most acute problem claiming attention is the situation in the south—in Zimbabwe and Namibia, where the South Africans still keep their troops quite illegally, as well as [the situation] in South Africa proper." He hardly had to expatiate on the Soviet Union's position concerning

25. Quoted in *Pravda*, October 2, 1977.

racism, the denial of the right of self-determination to the African nations, etc.

Gromyko's conciliatory tone was balm to the administration, which had already begun to be criticized for its excessive politeness to the Russians, especially in view of their brusque rejection of Washington's initiatives on SALT II. And now those in Washington who advocated patience when dealing with the Soviets could be encouraged when the redoubtable Gromyko himself conceded that "our two countries have entered upon a path which will lead to an agreement."

Gromyko held out a special bait for Carter. The President had made no secret of his eagerness for a summit meeting with Brezhnev. Despite long and on the whole not very encouraging experience of such encounters insofar as Soviet-West relations were concerned, Carter, like practically every American President since the war,[26] was confident that his amiable yet forthright personality would make just the right impression on the chief Soviet potentate: here was a man who was full of the friendliest feelings toward the USSR, but by the same token not one to be trifled with or bullied. And the picture of the two Presidents[27] beaming at each other would undoubtedly prove soothing to the world at large and probably bring some political dividends at home.

The Soviet foreign minister chose, though in a somewhat condescending manner, to be very encouraging about the prospects of a Brezhnev-Carter meeting: "The President [in his talk with Gromyko] did not conceal his desire to meet with L. I. Brezhnev. Besides, he has talked about it for some time. . . ."[28] But the question was, continued Gromyko, should such a meeting be preceded by careful preparation or should it just take place for its own sake without the two countries having come close to agreement on a number of issues prior to the

26. The only exception was Harry S. Truman, who after Potsdam wrote to his aged mother in Missouri that Stalin and Co. were simply unbearable and he hoped he would not have to see them again.

27. Brezhnev had just dumped unceremoniously the former official Soviet head of state, Nikolay Podgorny, and added this job to his other numerous dignities.

28. *Pravda*, October 2, 1977.

summit? Obviously the former. Then, undoubtedly, "it would prove to be a major step in the development of Soviet-American relations and a major achievement for peace." In other words, you can have your summit, but first you must meet us more than halfway on several controversial problems.

It may indeed have been the bait of a summit dangled before them which prompted President Carter and his advisers to agree to the Soviet initiative for reconvening the Geneva conference on the Middle East, of which the two powers were official co-chairmen, but which lapsed into abeyance shortly after it was convened following the 1973 war. In a joint communiqué the two countries specified that the conference should take up all the outstanding problems in Israeli-Arab relations, such as "evacuation by the Israeli troops of the territories occupied in the 1967 conflict [and] the Palestinian problem, including the securing of the rights of the Palestinian nation." [29] After these questions had been settled one could then take up the problem of safeguarding the frontiers between Israel and the Arab states by instituting demilitarized zones, international guarantees for the frontiers, and the like, with both the United States and the USSR participating as guarantors!

Great was the consternation in Tel Aviv and among Israel's friends in the US and elsewhere when the communiqué was published. The joint statement seemed to hint at the participation of the Palestine Liberation Organization in the talks, precisely the point that had led to the collapse of the Geneva conference in December 1973. Hitherto the Israeli government had stoutly denied the very existence of a Palestinian nation and was categorical in its opposition to a Palestinian state being erected on the territories seized from Jordan in 1967. The PLO for its part had rejected the notion of a Jewish state even within Israel's pre-1967 borders. Prime Minister Begin, after a tempestuous interview with the American ambassador to Israel, was hospitalized, allegedly because of exhaustion. The statement seemed to erode the previous US stand that it would not discuss the Middle Eastern question with any party denying Israel's right to exist.

29. Text of the communiqué in *Pravda*, October 2, 1977.

Nor was the joint statement received with rapture in Cairo. President Sadat and his advisers thought that the revived Geneva conference would lead to endless discussions and disputes, with the USSR acting as the cheerleader for the hardline Arab states. If by some miracle a provisional agreement could be reached, it would have to include an enhanced Soviet presence, perhaps even Soviet troops, in the Middle East.

It was clear that the US had fallen into a Soviet diplomatic trap. The joint statement "was a Russian idea and the final text was based on a Soviet draft."[30] The Carter Administration's spokesmen tried to calm the political furor aroused in Congress and elsewhere by stating that there was no question of the US negotiating with or recognizing the PLO as long as the latter refused to acknowledge Israel's right to exist. At the same time, confusingly, Carter in a press conference did not preclude the possibility of the PLO being represented in some way at Geneva, perhaps by its members being included among the Arab states' delegates. In any case, the President claimed, the statement did represent a Soviet concession to the American position: it marked the first time that the USSR had acknowledged the existence of Israel as a sovereign state. This led to further embarrassment, since the American press immediately pointed out that the Soviet Union had been the first UN member fully to recognize Israel in 1948 and had maintained diplomatic relations with it in the periods 1948–52 and 1953–67.

The administration's embarrassment must have been watched with amusement in Moscow, as was Washington's attempt to half-pressure and half-cajole Israel to participate in the ill-fated conference. The joint Israeli-American statement defining the groundwork for the proposed meeting was castigated by the Soviet press as an attempt to sabotage a peaceful solution for the Middle East. *Pravda* on October 25 quoted approvingly Syria's foreign minister, who stated that any resolution of the problem must be *preceded by Israeli evacuation of all* the occupied Arab territories and by firm guarantees of the rights of the Arab nation of Palestine.

30. Bernard Gwertzman in *New York Times*, October 7, 1977.

But Moscow's elation at its diplomatic coup turned out to be short lived. The necessity of avoiding the gathering in Geneva and its likely consequence drove Sadat to the drastic step of announcing that he was ready to go to Israel in person and initiate a dialogue with its leaders. This dramatic event took place on November 19, and was preceded and followed by a sulphurous Soviet press campaign against the Egyptian President. The erstwhile Soviet ally was now presented as an accomplice of the Zionists and a tool of the United States. The press quoted approvingly various Arab criticisms of the man who until then had been acclaimed as the hero of the 1973 war, the war that restored the Arabs' self-respect: how could the Egyptian President as much as set his foot in Jerusalem, the Arab part of which, sacred to the whole Moslem world, was being profaned by Israeli occupation? How could Sadat, the alleged champion of Arab unity, even hint at the possibility of a separate peace with the Israelis, shake hands and be entertained by those who had caused three million of his kith and kin to become homeless refugees? On its own *Pravda* dug into Sadat's past and came up with a sensational "revelation" (actually the facts were well known and had been stated in the Egyptian leader's autobiography): during World War II Sadat, along with a number of young Egyptian officers, had planned a coup against the British, then using his country as a base for their North African campaign against the Axis. Small wonder that a man who had once been willing to help Hitler has now become an agent of the Zionists and their imperialist allies! Possibly excepting the case of Mao, the Soviet press had never descended to such a level of vituperation with regard to the leader of a state with which the USSR still maintained formal relations.[31]

The Soviet diplomatic manoeuvre having thus misfired, the Geneva conference on the Middle East disappeared from the horizon. There could be one slight consolation for the Kremlin in the fact that the threat, as Sadat saw it, of the Russian interference with the peace process, drove him to seek a *rapprochement* with Israel more rapidly than he otherwise might

31. *Pravda*, December 7, 1977.

have planned, and thereby Egypt did find itself isolated within the Arab world. Even the more moderate Arab states such as Saudi Arabia and Jordan felt constrained to express disapproval of his precipitate trip to Israel, and, at least in public, proceeded to dissociate themselves from his initiative. But there was noticeable relief in Washington: the courageous Egyptian statesman had saved the Carter Administration from what undoubtedly would have been a major international, as well as domestic, embarrassment.

By the same token there could now be no question of an early summit meeting. After the US shifted its stand on the Middle East negotiations following the Sadat trip, the USSR saw no reason to appear overly cordial toward the Carter Administration or to "reward" it for what the Kremlin saw as a clear breach of America's commitment to the Geneva conference. In fact the Kremlin's innermost feelings about the administration probably paralleled the criticisms to which it was being subjected at home and by the US allies: it was virtually impossible to discern any clear-cut direction in US foreign policies, to assume that a decision or commitment announced by Washington one day would not be taken back some time later. There was obviously a considerable division of opinion in the President's entourage as to how to deal most effectively with the Russians. The Soviets, or at least their press, were quite explicit in identifying the main parties to the dispute and their arguments. The State Department, they believed counseled patience and perseverance in trying to find a common language with Moscow on at least some issues. The other group of officials, which, it was hinted included the national security adviser, Brzezinski, advocated a tougher course—for example, linking agreement on strategic arms to Soviet restraint in Africa, closer ties with Beijing so as to impress the USSR with the danger it incurred in continuing an expansionist course, etc. The President himself oscillated between these views, leaning to the "tough" side at one time, speaking and acting in the spirit of conciliation and hope at another.

A good example of this unsteady course of America's ship of state was provided by the question of the neutron bomb.

Despite American pledges to the contrary, SALT I had to cast doubts concerning the credibility of the US "nuclear umbrella" over Western Europe. For all their continuing support of detente, this country's European allies were becoming concerned over what they saw as the growing disproportion between the Soviets' conventional and tactical nuclear forces and those of NATO. One way of offsetting this disproportion lay in increasing the effectiveness of NATO's own short- and intermediate-range atomic weapons. It was in pursuance of this goal that in the summer of 1977 the Carter Administration requested funds for the development and manufacture of "enhanced radiation–reduced blast weapons," commonly known as the neutron bomb. As its description indicated, the bomb was supposed to work mostly through lethal radiation, rather than explosive power. Hence it was expected to be especially effective in countering or minimizing the superiority in tanks that the Warsaw Pact armies enjoyed over those of NATO. There was nothing about the bomb that made it more or less morally acceptable than any other frightful devices for destruction at the disposal of the two sides. Yet somehow— whether because of some misguided Madison Avenue specialist attached to the Pentagon or an inspired Soviet propagandist who characterized it as such—it became known as the bomb that killed people but did not unduly damage inanimate objects. In brief, an ideal weapon from the capitalist's point of view, since it annihilated people but left property intact! The Soviets, one hardly needs to add, were quick to seize upon this supposed "class" character of the weapon and to use it as the chief motif in their propaganda campaign to persuade American and European public opinion to demand the banning of this inhuman device. Orated Gromyko: "The struggle against the production and deployment of neutron bombs must be waged not only by us, the Soviet people. . . . [and] the peoples of the countries of the socialist community, but also by the people of each European state . . . whether it is or is not a member of this or that grouping."[32] And on January 23

32. Quoted in *Pravda*, January 4, 1978.

Brezhnev sent messages to the heads of NATO governments warning them against the introduction of the neutron bomb.

The intensity of the Soviet propaganda campaign reflected more than the realization that, by minimizing the effects of the Soviets' superiority in tanks, the bomb would make the Western European governments more confident that the Russians would not attempt aggression. There was also the Kremlin's shrewd psychological insight that anything which reminded the people in the West about the possibility of a nuclear war was bound to lead to widespread protests and pressures upon their own governments to negotiate with the USSR rather than install yet another horrendous weapon. And indeed the bomb, which was intended as a deterrent to an attack, rather than as an offensive weapon, and was potentially less destructive than many other arms long installed in Western Europe and European Russia, became the object of a heated and emotional debate, with many politicians arguing that its addition to the West's arsenal would make a nuclear holocaust more likely. No such heated discussion ever accompanied the launching of a US nuclear submarine, or the introduction of the Soviet SS-20 intermediate missile, each with destructive powers dwarfing those of the "capitalist bomb."

Several European statesmen, notably Chancellor Helmut Schmidt of West Germany, who in private had urged the US to proceed with the manufacture and deployment of the neutron bomb, found themselves constrained to state in public that they had gone along with the idea only because of the insistence of their senior ally. One may well imagine their discomfiture when all of a sudden President Carter announced on April 7 that he had ordered the production of the enhanced-radiation weapon deferred. The White House thus acceded to the pleas of those of President Carter's advisers who felt that this gesture by the US would in turn provide an incentive to the USSR to show similar restraint in deploying its own tactical nuclear arms in Europe. In fact nothing of the kind happened; the Soviets continued to modernize their tactical weapons, introducing more SS-20 mobile multiwarhead mis-

siles capable of striking any target on the Continent and in
Britain, and whose range could be considerably expanded with
some technical alterations. The President's indecisiveness, the
awkward position in which the announcement placed the
governments of America's allies, finally the continuing Soviet–Warsaw Pact military superiority in the European theatre,
all could not be displeasing to the Kremlin nor conducive to
greater restraint on its part, whether in respect to weapons or
other policies vis-à-vis the West.

In its attempt to loosen the Western alliance, the Soviet
Union steadily pursued tactics that antedated detente. Western Europe was to be persuaded of the Soviets' peaceful intentions while being at the same time impressed by Moscow's
growing military might. The average German, Dutchman, etc.,
was thus expected to feel that Moscow meant no harm to his
country, yet still be frightened by Soviet power—something
that might appear incongruous, but then logic is not an infal-
lible guide to political thinking. These mixed feelings would
then be translated into the pressure of public opinion upon
the Western governments not to tie themselves too closely to
the erratic policy of the United States so as to avoid provoking
the essentially peaceful (though dangerous if aroused) Soviet
colossus.

The overall situation had thus changed considerably from
that of the 1950s and early 1960s, when all Western European
countries saw close association with the United States as virtually the sole guarantee of their security, and some—e.g., Adenauer's Germany—in fact pleaded on occasion with Washington to adopt a harder line toward the USSR and not to put
much trust in the Kremlin's rhetoric about peaceful coexistence. By the late 1970s many in Western Europe, including
some politicians, had grown skeptical of the value of the Atlantic alliance and fearful lest a sudden shift in America's policies expose their own countries to new threats and dangers
from the Soviet side. The decline of faith in America's capacity for leadership of what had been known as the free world
(characteristically this term had now fallen into disuse, being
identified with the rhetoric of the Cold War) and the arrested

momentum for Western European unity were from the Soviet point of view positive developments on the international scene.

For its own part the Soviet government was sparing no efforts to convince the individual European states to emancipate themselves still further from American influence and to strengthen their bilateral relations—primarily in the economic sphere, but with a clear political undertone—with the USSR. Thus Soviet officialdom and the media gave great prominence to Brezhnev's state visit to the Federal Republic in May 1978. An impressive array of ministers and technical and economic experts accompanied the Soviet leader to Bonn. The two sides noted with satisfaction the growth in West German–Soviet trade, whose value had increased almost tenfold from the 1970 level: the Federal Republic in 1977 exported to the USSR goods worth 6500 million DM, while Soviet imports totaled 4500 million DM. An agreement signed during the visit projected a further expansion of economic cooperation between the two countries. Under the agreement German firms undertook to erect a number of industrial plants in the USSR. Much of that infusion of Western technology and machinery was to be financed by medium- and long-term credits, which Bonn promised to provide under relatively low rates of interest.

Though he was now flogging what was at least for the time being a dead horse, Brezhnev used the occasion for yet another public denunciation of the inhumanity of building the neutron bomb. It was a weapon so immoral that the Soviet Union would not allow the US to use it as a bargaining chip. The USSR, he said, would neither cut its own forces in Europe nor dismantle any of its new tactical missiles in return for US abstention from producing the bomb. His own bargaining position undermined by the recent American statement, Chancellor Schmidt could only complain to his guests about the growing strength and variety of weapons deployed by the Warsaw Pact forces along the borders of his country. But this dissonance on what might have seemed a cardinal point was not allowed to spoil the visit. In the concluding communiqué the two political leaders expressed their concurrence on a wide range of international issues, stressing the need "to expand

and deepen detente and to make it progressive and stable in character."[33] Though Brezhnev was not in his best physical shape, his visit was an undoubted success for Soviet diplomacy.[34] A special joint statement by the Politburo and the Soviet Council of Ministers hailed the visit as having "proceeded in a spirit of mutual understanding, frankness, and growing confidence."[35] It would be interesting to know whether Chancellor Schmidt privately would have endorsed the last part of that statement.

In any case, one could have hardly described the current state of Soviet-American relations as being characterized by a "spirit of mutual understanding . . . and growing confidence." The Kremlin's diplomatic offensive in Europe, combined with the Soviet-Cuban military activities in Africa, had placed Washington in an awkward position. America's allies, both singly and collectively, pressed for the continuation of the process of detente, mainly a speedy agreement on SALT II. At the same time the Soviets' and their allies' antics in the Third World, and the human rights situation in the USSR, made it increasingly difficult for the administration to justify to itself as well as to Congress and public opinion a conciliatory and hopeful attitude toward Moscow. SALT negotiations had resumed formally in May 1977, but there had been no signs of the USSR being willing to budge from its previously stated position on such controversial points as its heavy missiles, the Backfire bomber, and the cruise missiles. If the administration believed that by cancelling production of the neutron bomb it might affect favorably the SALT discussion, then, as we have already seen, the Soviets were emphatic in rejecting any such linkage. Even before giving up the neutron bomb, the US had discarded another potential bargaining chip (though in this case mainly for budgetary reasons) by discontinuing the manufacture of the B-1 bomber, originally designed to replace the aging B-52 as the main intercontinental

33. *Pravda*, May 7, 1978.

34. Embarrassingly, photographers happened to be present when he had to be helped to stand up.

35. *Pravda*, May 12, 1978.

heavy bomber. The ubiquity and complexity of the Soviet challenge would have baffled the most determined and single-minded democratic policy-makers, and as has been noted, the Carter Administration was far from being united on how to deal with this disturbing phenomenon. On the diplomatic front the US had the unenviable task of dealing with its allies' complaints that it was being both too hard on Moscow (in not proceeding more expeditiously with SALT II and in haranguing it about human rights) and too soft (in tolerating Soviet incursions in Africa and canceling the neutron bomb). And in turn Paris or Bonn, often resenting the Americans' preaching about the danger of their getting too close to the Russian bear, criticized US policies—e.g., in the Middle East—as being too rigid and thus offering new opportunities for the Soviets.

An equally incongruous chorus of criticisms, both foreign and domestic, accompanied Washington's efforts to stem the Soviet advance in the Third World. Some said the US was plainly encouraging the Russians by its supine attitude to what was happening in Angola, in the Horn of Africa . . . No, argued others, the Soviets' successes there, if indeed they were such, were mainly due to the Americans' insensitivity to the needs and feelings of the emerging nations, to their paranoid tendency to see the hand of Moscow in the new states' and liberation movements' natural strivings to remove the last vestiges of imperialism and white supremacy.

To a dispassionate analyst it must have often seemed that the conditions of the US-USSR rivalry in the Third World were almost inevitably weighted in the Soviets' favor. Political and social instability in the vast areas that had once been colonies or parts of the imperial spheres of influence was practically the normal condition; revolutions and coups d'etat were frequent. It was thus easy for Moscow to export, so to speak, matches to those vulnerable parts of the globe. And it was usually only when an actual fire had started that the American democracy would start deliberating whether and what kind of fire-fighting equipment should be dispatched to the threatened spot. A regional specialist in the State Department or National Security Council might spot portents of danger in some remote but crucial area, but a typical politician, not to mention your man in the street, would hardly be expected to become excited over the supposed "Communist danger" in a country he was barely able to locate on a map.

Thus it was that in April 1978 an event that in little more than a year would have profound international repercussions created but little stir in official circles in Washington and passed virtually unnoticed by the American public. A leftist coup in Afghanistan overthrew the five-year-old republican regime of Mohammed Daoud and established the rule of the Khalq (People's) Party, headed by Noor Mohammed Taraki. Western observers had no difficulty in recognizing the pro-Soviet character of the party and the regime, but were virtually unanimous in asserting that both the causes and the mechanics of the coup were purely internal. The former dictator Daoud, while accepting economic help impartially from both the East and the West, had been on very good terms with Moscow; many of Afghanistan's military and civilian officials had received their training in the USSR. As against this last fact, which Daoud probably came to regret during his final moments (he and his entire family were slaughtered in the uprising), Western analysts could draw some consolation from the American connections of both the number one and number two men in the new regime. Taraki had served in his country's embassy in Washington and then worked for the US International Development Agency and as a translator for the American Embassy in Kabul. His second-in-command and eventual nemesis, Hafizullah Amin, was an alumnus of Columbia University. It was also pointed out that of the twenty-one ministers appointed by the Revolutionary Council, "none had ever attended or been invited to attend Communist meetings."[36]

Seldom known to be sentimental in such matters, the Soviets, for all their past professions of friendship for Daoud, were quick to recognize the new regime. The influx of Russian military and civilian advisers, already considerable under Daoud,

36. Louis Dupree, "Afghanistan under Khalq," *Problems of Communism*, July–August 1979.

now became a veritable flood. In addition to concluding numerous trade and cooperation agreements with the USSR and
other Communist countries, the new government hastened to
enact legislation designed to transform the entire social and
economic life of Afghanistan. This last step could perhaps be
taken as an indication that as yet the regime was not entirely
under Moscow's thumb; the Soviets would hardly have advised such drastic measures, bound to stir up widespread popular resistance in a society barely emerging from the feudal-
tribal phase of development. That and the Afghans' fierce nationalist tradition (for more than a century the country had
resisted the efforts of two empires—the Russian and the British—to dominate it) led many outsiders to believe that the
USSR would not try to turn the new regime into its puppet.
Or, as a leading American expert wrote, the Afghans would
be allowed to keep "their internal political independence."[37]

Afghanistan's accession to the "camp of socialism" could
then be and was rationalized in this country on several
grounds: (1) Afghanistan was not important; (2) it had already
been under considerable Soviet influence before the coup; (3)
the upheaval was conceived and carried out without any outside intervention. Yet there remained the unpleasant fact that
another country had fallen prey to the Kremlin's long reach.
And on reflection, was Afghanistan so unimportant? It did have
a long border with Pakistan and had often laid claims to large
parts of the latter's territory.

In its earlier and more cautious phase the Brezhnev regime
might well have hesitated before giving its full blessing to what
had happened in Afghanistan. The purposes of Soviet foreign
policy in the region had been quite adequately served by the
Daoud government. Now Moscow assumed the burden of
dealing with the strife-ridden Kabul junta of self-proclaimed
Marxists, as well as that of protecting it against its own people, soon to be driven to desperation by the brutal attempt to
destroy their traditional beliefs and customs in the name of an
alien ideology. The events in Afghanistan, though in 1978 they
seemed to affect but little the overall world picture, were thus

37. *Ibid.*

a disturbing demonstration of the Kremlin's expansionist itch and its increasing confidence in the efficacy of its vast military power to offset any diplomatic and other costs and risks of collecting new satellites and dependents.

With its earlier optimism about mending US-Soviet relations fast vanishing, Washington was still at a loss as to how to counter Moscow's moves on the international chessboard or to make it realize that there were limits to American patience. Delicately at first, Carter sought to put the Russians on warning. In his Wake Forest speech of March 17 the President had declared it a myth "that this country somehow is pulling back from protecting its interests and friends around the world." And noting the Soviets' obduracy in the strategic arms negotiations, he commented that the US wanted an agreement, but not at any price: "Before I sign [one], I will make sure that it preserves the strategic balance . . . and that we will be at least as strong relative to the Soviet Union as we would be without any agreement." But this oratory was viewed by the Russians as directed mainly at the administration's domestic critics. And whatever beneficial effect such warnings may have had on the Kremlin, they were bound to be undercut by the subsequent American decision about the neutron bomb.

In May President Carter dispatched his national security adviser to China. Hitherto the administration had been quite circumspect in its dealings with Beijing, precisely for fear of ruffling Moscow's feelings. Now, with Brzezinski reported to be indulging in anti-Soviet jokes with his Chinese hosts, the Soviets could be expected to take American complaints more seriously. But it could not have escaped the Kremlin's notice that while there was now a lot of talk in the United States about playing the "Chinese card," nobody was quite sure how the card could or should be played. But anything touching on the possibility of closer links between China and the US was bound to make the Russians nervous. And there came now a certain stiffening in Carter's tone concerning the USSR. On June 7 at Annapolis he declared: "The Soviet Union can choose either confrontation or cooperation. The United States is adequately prepared to meet either choice." [38] Though the force of his challenge was somewhat weakened by the President's subsequent admission that when it came to nuclear weapons the USSR was superior to the US in several areas, the former having "more missile launchers, greater throw weight and more continental air defense capabilities," for once he used the kind of language that makes the Soviet leaders sit up and take notice.

Rather atypically, Soviet public commentary on the speech was not merely polemical in tone but reflected somewhat conflicting nuances in the Kremlin's appraisal of the President's tough talk. It was only on June 17 that *Pravda* published a lengthy commentary on Carter's speech; the delay in itself is indicative of how carefully it must have been analyzed. Why was Washington beginning to speak so harshly? Was it because of something the USSR had done? Obviously not. One possible explanation was that "for some time [America's] ruling circles have been engaged in an acute struggle over questions of detente and relations with the Soviet Union. . . . There are more and more indications that the forces gaining the upper hand in this struggle represent factions that would like to undermine detente and return the world to the Cold War." Yet there was a more hopeful possibility: that the President's strident rhetoric was meant to rebut the charge that he was being "too soft on the Russians" and thus to smooth the passage of SALT II when it came before the Senate. If so, he was surely taking a wrong approach. "The government itself is plainly in no hurry to take a definite stand and to begin to defend the agreement both in Congress and before the public. . . . On the contrary, many government leaders in this complex situation are busy stirring up mistrust toward the Soviet Union and spreading lies about 'the Soviet military threat.' " There were US officials who were still trying "to link disarmament talks to other questions that are totally unrelated." *Pravda* noted also "Washington's latest intrigues, or to be more exact, petty intrigues with China." Petty though they might be (and ob-

38. Quoted in *New York Times*, June 8, 1978.

viously the writer did not want the Americans to think they could get very far by playing the "China card"), they were still dangerous: "Soviet-American confrontation, and even better, war—that is Beijing's cherished dream."

The Russians, the article asserted, would not swerve from their course: "We do not join in the invitation to bury detente and the hopes of millions of people for a peaceful future." All in all, Moscow's mood was still one of cautious optimism, based on the convictions that the Carter Administration still needed and wanted SALT II, and that the American people were not ready to face a major international crisis; public opinion polls in the US indicated that the overwhelming majority of the people "favor the improvement of relations with the USSR, the achievement of an immediate strategic arms limitation agreement and the development of economic ties with the USSR." Hence this advice to Carter: "courage in politics is far from being the same thing as bravado and the readiness to resort to strong expressions and brandish a club. . . . By no means can you get away with the same things in foreign affairs as you can in American politics. . . . The price of miscalculation . . . will not only be measured by a drop in one's popularity. It already threatens to result in new and costly spirals in the arms race, political crises, a growth in the tax burden, and even more costly consequences." For all the growing strength of those elements in the administration and the country at large that desired a tougher stance vis-à-vis the Soviet Union, Moscow still believed that the President neither had decided upon nor could afford politically a drastic shift in American foreign policy.

At the same time it recognized that the growing American agitation over what the USSR was doing had to be countered by some propitiating gestures. On June 25 in Minsk Brezhnev reiterated a recent Soviet initiative in the Mutual Balanced Force Reduction (MBFR) talks dragging on in Vienna. He obliquely acknowledged the West's concern with the Warsaw Pact's preponderance in tanks. The Soviets proposed "The same common ceilings . . . for each of the groupings in Central Europe . . . , [which] corresponds to what the Western countries have proposed." If an agreement was reached the

Russians were ready to pull out three divisions and one thousand tanks within a year from the Warsaw Pact contingent.[39]

The Kremlin's consensus that the US policy-makers still wanted to avoid a drastic clash with the USSR was well founded. But while willing to give the Soviets the benefit of the doubt insofar as a SALT agreement was concerned and still eager for a personal meeting with Brezhnev, the President could not help irritating Moscow on other counts. The human-rights issue kept popping up, even though Carter had, since his first months in office, softened his rhetoric on the subject. But the US could not and did not remain silent in the face of what was seen in this country as the more flagrant violations of the Helsinki agreement. Soviet dissidents had publicly set up a group to monitor the regime's observance of the agreement, a step whose consequences its initiators must have expected. In May the group's former chairman, Yuri Orlov, received a stiff prison sentence for his allegedly anti-Soviet activities; the US and some other Western governments immediately voiced their indignation. In July and August there were further trials of human-rights activists. The one that attracted most attention abroad was that of Anatoli Shcharansky, accused, among other things, of passing secret information to an American journalist and sentenced to a ten-year prison term, to be followed by three years of exile. The contrived character of the charges was even more evident in the Shcharansky case than in other similar trials, the accused maintaining staunchly that he was being punished solely for advocating the right of Soviet Jews to emigrate to Israel. This time foreign protests were quite widespread, and as a tangible sign of America's displeasure the US government canceled its permission for the export of an advanced computer system to the USSR (which then without much trouble procured a very similar system from a French firm). The severity of the sentences against the dissenters and the rather clumsy attempts to tie their activities to the CIA, showed the determination of Soviet authorities to demonstrate to the Americans and others that they would not brook any interference with their domes-

39. Quoted in *Pravda*, June 26, 1978.

tic practices and that foreign protests were both futile and counterproductive. The vehemence of the Soviet reaction was not without effect. Carter hastened to explain that in his solicitude for human rights he did not mean to single out the Soviet Union, and that for the time being he planned no other punitive steps connected with its treatment of the dissidents. The Soviets evidently calculated that the American interest in the subject would have largely spent itself by the time the SALT II agreement was concluded and the treaty became the subject of public discussion during the Senate debate on its ratification.

Exasperating as the American criticisms of Soviet justice were to the Russian leaders, they still felt them to be but pinpricks, hardly of importance in the larger context of international politics. The US success in wooing Egypt away from the Soviet side was a different story. Until the last moment Moscow had hoped that the intractable nature of the Middle Eastern problem would block any agreement between Egypt and Israel, and it was difficult for the Soviets to swallow the fact that American diplomacy, usually so erratic and fumbling, this time proved capable of what had to be seen as its first solid achievement in several years: bringing together two hitherto irreconcilable enemies. But the primary object of the Kremlin's ire was not so much Carter as Anwar Sadat. Egypt's defection was seen to strike a blow at the whole rationale of Soviet policies in the Third World. Here was a country which for twenty years had been the pivot of the Russian penetration of the Middle East, recipient of billions in Soviet aid, on behalf of which the USSR (as its leaders probably managed to convince themselves) had risked on at least two occasions a confrontation with the West. If after all that one man could drastically alter the course of Egypt's policy, what kind of example would it set for other current allies of Moscow in Asia and Africa?

The Soviet official reaction to the Camp David agreement reached by Sadat and Menachem Begin on behalf on their respective countries was one of intemperate anger. Diplomacy would have dictated a more restrained response: at least a show of satisfaction that two countries formally at war since 1948 were able to terminate their abnormal situation. The USSR then could have stated with some reservations and regrets that the two sides had not addressed satisfactorily such key problems of the area as the rights of the Palestinians, the ending of Israeli occupation of *all* the territories seized after the 1967 war, etc. But the first Soviet commentary on the successful conclusion of the three-sided negotiations held in the President's Maryland retreat did not have a single positive thing to say about what had transpired there. The historic document signed by Carter, Sadat, and Begin was not even dignified by being called an "agreement." "A deal reached behind the back of the [Arab] nations" was the *Pravda* headline of September 19. Even before the first Arab reactions to the "deal" were reported, *Pravda* was certain that it represented a betrayal by Sadat not only of the cause of the Palestinians, but of the Arabs in general. "It is clear that the deal was reached at Sadat's expense." The Egyptian President was thus not only villainous, but a weak and naive politician, easily duped by the Israelis and their American protectors. The Israeli concessions envisaged in "the framework for a peace treaty between Egypt and Israel" signed at Camp David were proclaimed to be illusory and unlikely to be honored by Tel Aviv. In return for vague promises of returning the Sinai to Egypt, Sadat signed away the rights of the Palestinians and forfeited whatever sympathy and influence he had retained in the Arab world. No, this was not a step towards peace, "but a sellout transacted behind the back of the Arab nation, one which serves the interests of Israel, American imperialism, and the Arab reactionaries."[40] In fact the paper cited with satisfaction the negative responses to Camp David by those Arab states the Soviets have usually identified as the most reactionary ones: Saudi Arabia, Jordan, Kuwait. And the Soviet media made no secret of their hope that his own people would find ways of dealing with the man who isolated the Egyptians from their Arab brethren and struck an infamous deal with the Zionists.

40. *Pravda*, September 21, 1978.

The same bitterness over the Israeli-Egyptian accord, though expressed in a more restrained way, was evident in Brezhnev's personal references to it. Speaking in Baku on September 22 he said: "Already today it is obvious that the Arabs have resolutely condemned the separate deal in Camp David and turn away from it with indignation." After thus minimizing the significance of this setback for Moscow, the General Secretary seized immediately on what was on the contrary, in his words, "an international event of principal importance": Afghanistan. "There, as you know, has taken place a national revolution which overturned the semi-feudal regime. . . . We hail the Democratic Republic of Afghanistan—an independent, peace-loving, and nonaligned state—and are convinced that it will be a factor for peace and stability in that region of Asia. . . . We shall do all we can to strengthen our traditional friendship with this good neighbor."

In the same speech Brezhnev expressed Soviet impatience over SALT II negotiations. "Obviously the gap between the two sides' negotiating positions is not so great; it can readily be bridged with real goodwill and statesmanship." Some influential circles in America evidently did not wish "a stable peace and mutual cooperation." But for all such difficulties and the slanderous campaign about the alleged violations of human rights in the USSR, that country would persevere in its efforts to reach an agreement with the US. "We shall resolutely rebuff any attacks on the rights and interests of the Soviet state, but we shall not let ourselves be provoked." [41]

By the end of the year the Soviets' perseverance appeared to have won the day: SALT II was *almost* ready for signing, and on many of the originally contentious issues the draft treaty reflected Moscow's wishes. On the other hand there were increasing indications that once signed, the treaty would be in for rough going in the Senate, where it had to be approved by a two-thirds majority. Many senators had been impressed by the argument of the proposed agreement's opponents that its provisions could enable the USSR to launch a preemptive

41. Quoted in *Pravda*, September 23, 1978.

strike sometime in the 1980s that would destroy virtually all of the US land-based missiles and leave the country with the frightful option of either capitulating or inviting a Soviet second strike, which would level American urban centers, with casualties on the order of 100 million. Unlike those of the US, the Soviet Union's leaders, it was argued, did not adhere to the theory of mutual deterrence. They were preparing for and believed they could win a nuclear war.[42] The administration stoutly disputed such scenarios and allegations. In November, talking with some US senators visiting Moscow (an occasion probably harder on the Soviet leaders' nerves than almost any other aspect of detente) Brezhnev himself sought to refute the insinuation: no, the USSR did not think it could "win" a nuclear war, it would be an utter disaster for both sides and mankind in general.

In any case, Carter and his advisers believed that the agreement would in fact be in the best interests of this country, and that public opinion, after the inevitable debate, would see it as such, and as a major step on the road to real peace. The final phase of negotiations, with both Vance and Gromyko present, was taking place in Geneva—some rather minor technical details remained to be smoothed out. The treaty was to be ready, appropriately enough, by Christmas: ". . . in early December senior administration officials had leaked the word to the press that plans were under way for Brezhnev to come to Washington for a SALT II signing the week of January 15."[43]

It was at this point that Carter chose to announce that the US would establish full diplomatic relations with the People's Republic, withdrawing its recognition of as well as terminating its defense pact with Taiwan, and that China's current top leader, Deng Xiaoping, would visit Washington at the end of January. The special timing of the announcement had been urged on the President by a group of advisers headed by se-

42. The classical statement of this position is contained in Richard Pipes, "Why the Soviet Union Thinks It Could Fight and Win a Nuclear War," *Commentary*, July 1, 1977.

43. Strobe Talbot, *Endgame: The Inside Story of SALT II* (New York, 1979), p. 229.

curity assistant Brzezinski. They obviously hoped that this dramatic display of the China card would make the Soviets more amenable to the last-minute adjustments, admittedly not of major importance, in SALT II, as well as facilitate ratification by the Senate. Their arguments prevailed over those of Secretary Vance, who had been apprised of the President's decision only a few hours before its announcement. What Vance feared was also anticipated by the Chinese Communists, but on their part with considerable satisfaction: the President's move would infuriate the Russians, and at the least delay SALT II.

Indeed it was unreasonable, to put it mildly, to expect the Kremlin to go along with the White House scenario: impressed by the American diplomatic coup, the Soviets hasten to sign the agreement, and Brezhnev hurries to Washington to mollify the US just before the impending visit of Deng. The Geneva negotiations were broken off by the Russians, allegedly because of some technicalities that required further discussion. Instead of SALT II, Carter's Christmas present from Brezhnev was a personal message concerning the recognition of Beijing. Trying to put the best appearance on a somewhat troublesome situation, the President declared that Brezhnev's letter took a very positive view of the establishment of formal ties between the two countries. The extent of the Kremlin's pique may be judged from the fact that it then took the rather unusual step of releasing the gist of the Brezhnev message to the public to demonstrate that Carter was, well, not putting the correct interpretation upon what the Soviet leader had said. No, it was not a message congratulating the US for having finally recognized officially a fellow Communist regime with which the USSR had had diplomatic relations for thirty years (not to mention the Sino-Soviet Treaty of Friendship and Alliance, theoretically—very theoretically—still in force). Brezhnev expressed his unhappiness over the language of the joint Beijing-Washington statement, which said that the two countries were opposed to any power seeking "hegemony" in Asia and elsewhere. Whom could they have in mind? The letter then went on to state the obvious, but in terms that might be thought both insulting and menacing when addressed to the head of a world power: "The Soviet Union will most closely follow what the development of American-Chinese relations will be in practice and from this will draw appropriate conclusions for Soviet policy."[44]

A short time before, George Arbatov had spoken more explicitly about Soviet fears concerning US-China relations: "If China becomes some sort of military ally of the West, even an informal ally, . . . we would have to reexamine our relationship with the West. . . . Then there would be no place for detente." Arbatov then had some words of advice for Beijing, rather similar in their general thrust to what many Americans would have urged upon the USSR: instead of throwing its weight around and creating trouble all over the world, the People's Republic of China should "turn its efforts toward the tremendous internal problems they have developing the country and improving its standard of living, etc."[45]

One can only speculate what would have happened had SALT II been signed in December 1978 rather than June 1979. But quite apart from any assessment of the ill-fated treaty, the failure of the two governments to get together on something they both (though perhaps for different reasons) very much wanted had disturbing undertones. It was not atypical for the US to misjudge the psychology of the men in the Kremlin, but it was unusual for the latter to allow their ruffled pride to prevail over rational criteria that urged them to conclude the SALT II process, China or no China. These were not good portents for the future.

44. Quoted in New York Times, December 22, 1978.

45. Interview with Jonathan Power, International Herald Tribune (Paris), November 11–12, 1978.

5

Of Weapons and Hostages

While few of today's analysts of Soviet foreign policy would adhere to the once-fashionable view that its main source was to be found in the messianic urge to conquer the earth for Communism, many still hold world domination to be Moscow's ultimate aim, though for reasons having little to do with ideology. But even the proponents of the latter view would seldom accuse the Kremlin regime of recklessness in the pursuit of that goal: faced with insuperable obstacles, confronted by what they see as unacceptable risks, the Soviet leaders know how to pull back, change their strategy, and wait. For their own part, Russia's policy-makers, while rejecting as slanderous the imputation that they seek world mastery, would concur in the view that theirs is a prudent and realistic approach towards day-to-day international developments. And though, beginning with the mid-1970s, caution does not appear to us to have been the outstanding characteristic of Soviet policies, the Kremlin evidently still believed it could gauge correctly the opportunities and dangers inherent in its course, and preserve the right balance between expansion of Soviet power and influence on the one hand, and detente on the other.

Yet, by the end of the decade, one had to wonder whether it lay within the ability of any state, no matter how great its power and how skillful its diplomacy, to prevent the ever more widespread international conflicts and tensions from erupting eventually into an irreparable clash. Even the Soviets grew visibly less confident that they could anticipate and control every dangerous contingency on the international horizon. *Pravda,* in the already quoted editorial of June 26, 1978, responding to Carter's Annapolis speech, commented: "As we already see today, the 'tough line' has every chance of changing from a tactic into a dangerous and uncontrollable political course, acquiring a momentum of its own that would be difficult to overcome and would call forth an appropriate counteraction." This was quite prescient, but the writer lacked enough introspection to acknowledge how far certain of the Soviet Union's policies had already developed such a dangerous momentum of their own, and how in the near future they were bound to call forth "counteraction," whether appropriate or not, from the US and China.

One such momentum was that of what might be called secondary Soviet expansionism. The Afghanistan coup of April 1978 may well have been locally inspired and planned, but the fact was that the USSR hastened to throw its protective mantle over the new regime, and treat its preservation and continued and growing dependence on Moscow as a political imperative. And it became evident very early that the new rulers, deeply divided among themselves and having alienated the great majority of the Afghan people by their doctrinaire reforms, would be unable if left to their own devices to stem the rising tide of rebellion against their rule.

An even more complex and dangerous situation arose in Indochina during 1978. One could hardly consider Hanoi a mere puppet of the Soviet Union, but it was undoubtedly the latter's protection—and quite possibly its actual encouragement—that had emboldened Vietnam to assume an increasingly defiant stand towards its huge neighbor, the People's Republic. Territorial disputes and border clashes between these two once very friendly regimes grew increasingly sharp. And while neither side was blameless in the quarrel, it was Vietnam that undertook what in Beijing had to be seen as a clear provocation: systematic persecution and mass expulsion of its citizens of Chinese extraction.

Having seen its influence entirely eradicated in Vietnam and Laos, China could derive some consolation from its close ties with the Khmer Rouge regime in Cambodia—a regime that could not be viewed with much sympathy by Moscow or Hanoi or, for that matter, the rest of the world: during its brief rule, it gained the unenviable distinction of dealing more cruelly with its own people than any other dictatorship in modern history, whether of the left or right. Some strange ideological quirk led the rulers to drive most of the urban population into the countryside, to carry out wholesale massacres at the slightest appearance of resistance, and to subject the traditionally peaceable nation to a kind of doctrinaire regimentation that made Stalin's forcible collectivization of the 1930s look like a moderate reform. For once the Soviet press did not indulge in propagandistic exaggeration when it described what was going on in Cambodia as virtual genocide.

Still one suspects it was not humanitarian considerations but its pro-Chinese orientation that determined Hanoi, with Moscow's blessing, to liquidate the Cambodian regime. There were the usual preliminaries: border fighting, incitements to the Cambodians to rise against their oppressors, and, on December 3, 1978, the setting up of a Cambodian "National United Front for National Salvation," which the bloodstained regime denounced as a Vietnamese-Soviet puppet. There followed the invasion of Cambodia by Hanoi's army, which in less than a month succeeded in placing most of the unfortunate country under new masters. On January 9, 1979, the new government—an emanation of the "United Front"—was recognized by the USSR and its allies. Partisans of the old regime, characterized by the Soviet press (from the names of its two main figures) as "the Pol Pot–Ieng Sary clique," sought refuge in the more inaccessible and border regions of the country, from which to this day they conduct guerrilla activities against the Vietnam-imposed government.

The fact of aggression was undeniable, and despite the hideous record of the fallen regime, the UN refused to recognize and seat the new authorities. Even within the Soviet camp, Romania—its occasionally maverick member—condemned the Vietnamese aggressors.

The USSR did not even try to conceal its satisfaction over the turn of events in Indochina. The issue in Moscow's eyes clearly transcended Cambodia. The Pol Pot regime's defeat was a damaging blow to Beijing's prestige just at a moment when it had been giving itself the airs of a great power and Deng was setting out for Washington. The Soviet representative on the Security Council was at pains to elaborate this theme in his speech before the UN: "It is obvious that the Pol Pot–Ieng Sary regime would not have embarked upon the path of genocide of its own people by itself and without foreign support. The reason was that this regime was but a puppet of the external forces that are conducting the policy of hegemonism, great power chauvinism, and expansion in Indochina, as well as in Asia as a whole." [1] This was then a vivid demonstration of the value of the "China card," on which some Americans placed so many hopes: Beijing was impotent to save one of its very few friends in Asia, while the Soviet Union's protégé, Vietnam, was now master of all Indochina.

Both the Afghanistan developments of April 1978 and the Cambodian ones of January 1979 were hailed by Moscow as "contributing to the peace and stability" of the respective regions of the world; this was now almost a liturgical phrase when celebrating successes of Soviet foreign policy. Yet even some Soviet officials must have realized that in fact the very opposite was the case: the momentum of the Soviet Union's expansion was causing the Kremlin to incur commitments whose ultimate consequences could not be gauged or fully controlled, and which were not likely to contribute to peace and international stability.

Jubilation over the coup in Cambodia was soon tempered by fresh apprehensions. This was a clear provocation to China, and how would China respond? Simultaneously with its game in Indochina, Moscow sought to draw the West's attention to

1. Quoted in *Pravda*, January 13, 1979.

the dangers inherent in trying to get too close to the "Maoist clique" (Mao's death did not stop the Soviets from thus identifying the current rulers of China, perhaps a subtle hint that a really new generation of leaders in Beijing might again seek Russian friendship). Rather amusingly, in view of what the Russians themselves were doing at the moment, a Soviet commentator sought to warn Washington that by its *rapprochment* with China, the US was in effect giving the latter carte blanche to try to impose its rule upon Southeast Asia. This message was undoubtedly being echoed by the Soviet diplomats in Western capitals. The writer then proceeded to expose Beijing's other sinister designs. It was seeking to make Chinese influence dominant in the Arab world and the Persian Gulf(!). More than that, the "Maoist clique" clearly sought hegemony over the whole Third World and was greedy to exploit its resources for the "modernization" of its country—a code word for building up China's armed might.[2] There is no Russian equivalent for "the pot calling the kettle black"; our writer went on, blithely pointing out how short-sighted it would be for the West to help China with such "modernization." Alas, there were people in the US who were in favor of arming this expansionist regime, and while the Carter Administration was hypocritically denying any intention of selling offensive weapons to Beijing, it actively encouraged America's allies to do so. Hopefully (this now for the benefit of the Soviet reader, lest he become too alarmed at the prospect of an 800-million-strong giant equipped with the most modern weapons and military technology), the people of the US were less short-sighted than their leaders. "The policy of encouraging the Maoists in their expansionist strivings is clearly unpopular with the great mass of the American population." He then goes on to quote approvingly various voices in the West warning against the "Maoists' " inherent duplicity, about the danger of alienating the Russians for the sake of the perfidious Chinese, etc. And he concludes by comparing the heirs of Mao to Hitler and his gang.

2. Vitali Korionov, "The Short-Sighted Gamble," *Pravda*, December 25, 1978.

With Deng already in Washington, the Soviet media hastened to reveal alleged secret instructions from Beijing to its foreign missions, explaining the Sino-American *rapprochement* as a ruse designed to enlist the capitalists' help in restoring China's economy and thus enabling it "to strengthen itself and to achieve economic victory over both capitalism and 'social imperialism' [i.e., Russia]."[3]

Long experience should have taught the Russians the futility of trying to outshout their erstwhile Chinese friends. Somewhat to the embarrassment of his hosts, Deng Xiaoping, once in Washington, did not curtail his anti-Soviet rhetoric. The Soviet press seethed as it reported his public statements, in which he warned the West against trusting Moscow on any international issue, expressed his regime's skepticism about SALT II, and called for a US-China joint endeavor to "check the ambitions of the 'polar bear' [the USSR]." The boisterous Chinese leader, whose personality, if not exactly his views, enchanted American official circles and the media, was uninhibited in denouncing the recent developments in Ethiopia, Afghanistan, and Africa and in ascribing them mainly to Soviet instigation. He was equally frank, or indiscreet, in implying that President Carter's optimism about Moscow's ultimate intentions was unfounded. "Deng did not find a single word to say about socialism, the peoples' struggle for national liberation, and [other] revolutionary strivings in the world," noted *Pravda* indignantly.[4] There was some comfort for the Kremlin in the fact that the White House made it clear that it did not endorse its guest's more extreme pronouncements and that America's policy of not furnishing China with offensive weapons still remained in force.

In his private communications with American officials, Deng made no secret of the fact that his regime planned "to teach Vietnam a lesson" for its invasion of Cambodia. Even without his confidences it must have been obvious to both the Soviet and American intelligence services that China was planning a

3. *Pravda,* January 31, 1979.

4. A. Petrov, "Concerning an Interview by Deng Xiaoping," *Pravda,* February 1, 1979.

major military move against its former friend and protégé.
Beijing's ponderous and antiquated military machine was incapable of camouflaging its preparation. On January 30 *Pravda*
quoted an unidentified US State Department official who had
confided to *Newsweek* that the massing of Chinese troops on
the Vietnam border was assuming disquieting proportions,
adding, "We are worried that an invasion of Vietnam may take
place following Deng's visit to the US and thus give the
impression that we have given it a green light." [5] But neither
such indiscretions nor the American officials' arguments could
persuade Deng and his colleagues to call off the venture. The
ill-advised invasion was begun by the Chinese on February
17, their troops crossing into Vietnam all along the border.
This diffusion of military power, rather than concentrating the
attack on a few especially vulnerable points, was proof in itself of the primitive state of Beijing's strategic doctrine. Because of its great superiority in manpower, the People's Liberation Army was able eventually to occupy and devastate a
number of frontier posts and towns, but in military and political terms little more than two weeks' fighting resulted in a
considerable setback for the Chinese. They faced an enemy
much better equipped, organized, and led than their own army,
which still followed the tactics and logistics of the Korean War,
almost thirty years before. The USSR confined its own moves
to warning Beijing to terminate the invasion or face unspeci-
fied grave consequences, and to shipping military supplies to
Hanoi. The attack failed to compel Vietnam to reduce appre-
ciably its occupation force in Cambodia. Washington added
its plea to Moscow's threats: continuation of the invasion was
pregnant with incalculable dangers. The Carter Administration was sorely troubled by recent developments in Iran; it
certainly did not wish for yet another explosive international
situation to confront it with impossible alternatives. The Soviets, for their part, would not—at least in public—absolve
America from a large share of responsibility for what had been
happening. An article in *Pravda* signed by a pen name that
customarily stood for an authoritative Kremlin spokesman as-

5. *Pravda*, January 30, 1979.

serted that "The ambivalent position of the ruling circles of the US in the face of China's blunt threats against Vietnam at the very least contributed to Beijing's embarking upon open aggression."[6]

On March 5 the Chinese government announced that, "having fulfilled their assigned tasks," its troops were evacuating Vietnamese territory. Beijing suffered quite a blow to its prestige, even if that of the USSR was not exactly enhanced. On the Chinese side, losses, even by its own admission, were staggering: 20,000 killed in sixteen days' fighting. (By comparison, American deaths for the *whole* of the Vietnamese war had been about 50,000.) And if there was a lesson to be drawn from the affair, it was not of the kind that Deng and his colleagues had intended. On the other side there was the undeniable fact that, though its ally had been attacked, the USSR confined itself mainly to ominous warnings; quite a few Asians must have wondered what, if anything, Moscow would have done if China's troops had in fact reached the outskirts of Hanoi. Yet the Kremlin was not slow to claim a considerable share of the credit for its enemy's humiliation. "And now the Chinese leaders have been forced to beat a retreat. This has come as the result of the steadfastness and courage of the Vietnamese nation and [because of] military readiness on the part of the Soviet Union . . . and other states of the socialist alliance,"[7] said Brezhnev in a speech of March 13.

Few could believe that the end of the invasion spelled the beginning of peaceful coexistence between the two Asian Communist regimes, once so closely linked. Border clashes and mutual vituperation would go on, and the intermittent negotiations to settle the issues allegedly in dispute between Hanoi and Beijing have paralleled in their futility those between Moscow and the People's Republic. And indeed the melancholy story of Sino-Vietnamese relations has been yet another lesson on how a common ideological heritage has frequently served to aggravate rather than dissolve traditional national animosities. A Kuomintang China would have undoubtedly gotten

6. *Pravda*, February 28, 1979.

7. L. I. Brezhnev, *Following Lenin's Path*, vol. 7 (Moscow, 1979), p. 633.

along better with the Soviet Union than has that of Mao and his successors. Beijing felt, and with a great deal of reason, that the very survival of Communism in Indochina, not to mention its victories over the French and the US, would have been impossible without the help the Chinese Communists had rendered at a time when the USSR showed but little interest in Vietnam and the rest of the area. And now Vietnam had become a bastion of Soviet influence in Southeast Asia, and an enemy of its erstwhile saviors.

All of the Kremlin's gloating over China's discomfiture could not, however, change the fact of the latter's huge population and resources. To be sure, after Mao's death there came revelations baring the numerous weaknesses and vulnerabilities of the state and society he had ruled for over a generation. The Cultural Revolution had already shaken the image of a nation united around its leader, with state and party machinery free from factional strife. That image had been quite widespread in the West, where many, by no means only on the left, had come to believe that China had eliminated poverty so that for the first time in its history the peasant masses were being adequately fed, and that, unlike the USSR, it had a truly egalitarian system, with no privileged bureaucracy or other departures from the ideological code. Now the picture as revealed by Mao's successors appeared quite different. By Beijing's own admission, 20 percent of the country's collective farms were unable to feed themselves, and the industrial workers' wages had not been raised during the last two decades of Mao's reign. All such avowals were being eagerly noted and emphasized by the Soviet press.[8] And how could it be different, the Soviet reader was told, when the Maoist clique had dissipated the country's resources for the sake of its "adventurist policies" abroad, spurned Soviet help and advice, and indulged in doctrinaire and ruinous economic practices?

But such a by and large realistic depiction of China's troubles was usually followed by a dubious prescription for their cure: Beijing should rejoin the socialist camp and resume close

8. E.g., the article on the Third Plenum of the Communist Party of China, *Pravda*, December 28, 1978.

ties with the USSR. Had the Kremlin been really serious about regaining China's friendship, it certainly would have played a different game in Indochina. The real hopes of Moscow concerning the People's Republic still centered on the expectation that the factional strife within its Communist Party would grow ever sharper, perhaps eventually assuming the proportion of a civil war. Under such conditions one of the contending factions might seek Soviet help. Yet for all the real and serious cleavages within China's ruling group, even the authors of the Kremlin's scenario must intermittently have realized that it was only wishful thinking. Chinese distrust of the USSR was not due to some malevolent contrivance by the "Maoist clique"; at first latent, it had during the previous two decades become deeply ingrained among all factions of the Chinese Communist Party. And, by the same token, no foreseeable papering-over of the Moscow-Beijing dispute was ever likely to make most Russians lose their fear of China, no matter who its rulers were.

It is against this background that one must see Brezhnev and his colleagues' approach to the Chinese dilemma. For all their professions of the possibility and desirability of reconciliation, their actual policies sought to isolate the People's Republic. If left entirely to their own devices, the Chinese Communists, the Kremlin believed, would need several decades to build their country into a first-class industrial and hence military power, and only then could China pose a mortal danger to the USSR. But the process of industrialization would be considerably speeded by massive help from abroad, i.e., from the West and/or Japan.

Both detente and the massive Soviet arms build-up were then intended, among other things, to deter the West from offering such help to China. Detente was to persuade the US and its allies that they had much more to gain, both politically and economically, by being friends with Russia rather than China. And the arms build-up was to complement Moscow's entreaties with an implicit threat: think of the dangers you would face in incurring Soviet hostility by being too helpful to China! The poor and backward People's Republic was by comparison of little use as a friend, and much less frightening as a potential enemy.

The brutal logic of such arguments could not be ignored by Western policy-makers. Even as they entertained Deng, US officials were confronted by yet another major international crisis, one in which China could be of little help to them, while if the USSR chose to, it could raise the crisis to the level of confrontation.

After tottering for well over a year, the imperial regime in Iran finally crumbled, and the Shah left the country on January 16, 1979. "The Shah's departure ushers in a new stage of the Iranian crisis. Those Washington gamblers who indulge in hazardous games have suffered a humiliating defeat. Was it not just very recently that Brzezinski asserted that the Shah is supported [by the US] one hundred percent, and unconditionally, and that the White House gave assurances it would do everything possible . . . [and] would stick with the Shah come what may? And now that game has fizzled," rejoiced the Soviet press.[9] The Soviets could not deny themselves the satisfaction of noting how the very same people who had thought they could bluff them with the "China card" were thrown first into utter confusion, then into something resembling panic by the recent developments in Iran. This jubilation was not allowed to interfere with the Kremlin's vigilance: for all their agonizing indecision, the Americans, it was insinuated, might still try some last-minute tricks to save vestiges of their influence in the oil-rich country. "The foreign press has relayed disquieting news from Iran. . . . The country is threatened by a pro-imperialist conspiracy, which would set up a dictatorship. What is being envisaged is rule by the military, supported by the US military-industrial complex and instructed by its special espionage agencies."[10] And in support of such charges, Pravda quoted Khomeini saying that "The conspirators are seeking a pretext for setting up a dictatorship and for wreaking vengeance on the people." As happens not infrequently, the American press, with its masochistic reflections

9. Pravda, January 29, 1979.
10. Pravda, January 20, 1979.

on yet another debacle of US foreign policy, provided good
copy for *Pravda:* "We have done in Iran the worst we could.
We returned the Shah to his throne to save the country for
[our] oil companies, we filled Iran with CIA agents, instructed
its secret police, sold [the emperor] arms he did not need and
sent there thousands of American advisers," was a fairly typical lament of a US journal,[11] this one overlooking the fact that
Iran's oil fields and refining facilities were state-owned.

But Moscow's and some Americans' fears were soon dissipated: after a period of agonizing indecision, Washington advised the Iranian military not to attempt a coup. Khomeini
returned in triumph and initiated that peculiar regime of semi-
dictatorship and semi-anarchy which, in terms of the terror
and havoc it wreaked on the unfortunate country, was to surpass the Shah at his worst.

Apart from its anti-Western character, there seemed few reasons for Moscow to rejoice so demonstratively over the Iranian
Revolution. The Shah's attitude toward the USSR had been
more than correct from the latter's point of view. When a Soviet pilot defected to Iran, both he and the plane were returned to the Soviets forthwith. Iran's natural gas had been
piped into the Caucasus, allowing its own energy resources to
be profitably exported by the Soviet Union. When it came to
the emperor's relations with the US, it might be reasonably
argued that rather than being their puppet, he had in fact used
the Americans. Thus Iran had been one of the most strenuous
advocates of OPEC's quadrupling of oil prices in 1973–74, and
a staunch proponent of full exploitation of the organization's
monopoly position. As long as he seemed to be firmly in
power, Mohammed Reza was treated with deference and even
courted by the USSR, invited officially and feted in Moscow,
as well as in other capitals of the Communist bloc.

Once the Shah's fortunes began to decline, the Soviets' attitude toward him underwent a perceptible change, and when
the possibility of his being overthrown changed into a probability, Moscow's tone turned to vituperation. The once-honored
guest of the Soviet government now became "America's po-

---

11. Quoted in *ibid.*

liceman in the Persian Gulf"; there were repeated references to the opprobrious activities of SAVAK, Iran's secret police. The Soviets' now quite explicit support for the anti-regime forces was uninhibited by the fact that the leading group among them was the extreme Moslem fundamentalists, i.e., essentially the same element currently causing so much trouble for the Moscow-sponsored government of Afghanistan. But what in Afghanistan was a reactionary and anti-popular force became in the Iranian context a patriotic and progressive movement: "There is nothing unusual in the fact that the struggle against the [imperial] regime has been led by religious activists. The peculiarity of the Iranian situation consists in the fact that the majority of the population there is under the influence of the Shiite branch of Islam, whose slogans under the given circumstances have a progressive character. . . . [The slogans] call for the people's struggle against the monarchy and foreign oppression and for heeding the social needs of the masses." [12] And to convince a skeptic who might wince at this inclusion of the Ayatollah in the hall of fame of progressive leaders of mankind, the writer felt constrained to appeal to the ultimate authority: "In his time, V. I. Lenin taught that in the Orient the national movements not infrequently assume some peculiar forms."

Though, or possibly because, the chances of American military intervention in the Iranian imbroglio following the Shah's flight were nil, the Soviets warned Washington against such an eventuality. The USSR drew attention to a provision of the Soviet-Persian treaty signed after World War I which, while renouncing special rights Tsarist Russia had usurped in northern Iran, authorized the reentry of Russian troops in case a third power interfered militarily in the country's affairs. [13]

Treaty or no treaty, it was obviously in the Kremlin's interest to let the Iranian crisis ripen and then fester by itelf, rather

12. *Pravda*, January 24, 1979.

13. In fact that treaty had been superseded by the Big Three agreement during the Teheran conference of 1943, which pledged the US, the USSR, and Britain to withdraw their troops from Iran within six months of the end of World War II and to observe unconditionally the country's sovereignty.

than to send in Soviet troops, a step which for all of Washington's paralyzing indecision would have almost inevitably led to some kind of American response. The new Teheran regime was strong enough to embark upon a wholesale massacre of the former emperor's partisans, especially within the military, but seemed barely able to control the mass turbulence that had toppled the Shah, and still less to keep the country's economy on an even keel. Iran's oil production and exports had catastrophically declined, which, apart from dealing a blow to the country's finances, had the effect—not unwelcome from the Soviet point of view—of compounding the West's economic troubles.

The overall situation thus opened to Moscow a whole vista of alluring possibilities. The Revolution enabled the pro-Soviet Tudeh Party to come out into the open. For the moment it joined in supporting Khomeini, but if the religious autocracy faltered or fell prey to factional strife, the Iranian Communists would become serious contenders for power. Political instability and economic distress strengthened separatist tendencies within the multiethnic state. When occupying northern Iran during and immediately following World War II, the Soviets had sponsored Azerbaijani and Kurdish secessionist movements, both of which eventually proclaimed "independent" republics. These disappeared very soon after the USSR, under strong US and British pressure, finally removed its troops from the area. Should the occasion warrant it, the Kremlin could replay its 1945 gambit.

Moslem fundamentalism, the dominant force in the Iranian Revolution, was in principle as anti-Communist as it was anti-Western. In fact, however, it was the US that became the primary target of attacks by the mullahs and by fanatical crowds, primarily because of the pervasiveness of the American presence under the Shah, but also, one suspects, because even the fanatics found it more prudent to heap imprecations on the distant US than on the USSR, Iran's neighbor.

Would not the anti-Western fury and obsession soon spend itself, and the more moderate elements in the coalition that had overthrown the Shah succeed in steering the new regime towards a more balanced attitude in its international outlook? The Carter Administration hoped that after the tumult and shouting had died down, practical considerations would make the men now in power view the US not as the "great Satan" of Khomeini's rhetoric, but as a desirable commercial partner, the provider of spare parts for Iran's industrial and military equipment (acquired mostly from America), etc. But rational considerations are not always decisive in revolutions, especially those of the Iranian kind. Just in case, however, Soviet propaganda, especially through broadcasts beamed from radio stations in the Caucasus and Central Asia, kept expatiating for the benefit of Iranian listeners on the evils of American imperialism, its collusion with the Shah and the Zionists, and the nefarious designs Washington still entertained concerning their country.

US complaints about the Soviet media's anti-American propaganda were answered by Moscow with a mixture of indignation and facetiousness. Washington would have one believe, wrote *Pravda,* "that it is none other than the Soviet press and radio that are responsible for the recent developments in Iran and the subsequent discomfitures suffered there by the Americans." [14] It was presumptuous for a State Department spokesman to voice such complaints at a press conference and to assert that what the Soviets were writing and saying on the subject "could not but influence [negatively] the climate of detente." If the State Department was so solicitous about detente, why had it not reacted accordingly to the vicious anti-Soviet propaganda unleashed by Deng while in the US as a guest of the White House? No matter what the issue under discussion, it was always China that surfaced sooner or later as the focus of the Kremlin's concern.

Both superpowers sought and did find comfort in the other's troubles with its erstwhile allies. But the Soviet Union's difficulties on this count, while in the long run infinitely more serious, were for the moment much less pressing: China, as its setback in Vietnam demonstrated, was being contained. The Iranian crisis, on the other hand, led to another breach in that

14. February 18, 1979.

system of alliances by which Washington had sought to contain the USSR as well as to assure a degree of stability in an area which next to Europe was most crucial to the security of the West. To be sure the Shah, as we have seen, had not been a very reliable ally. The OPEC coup of 1973–74, which prodigiously enriched its members (and at the same time helped to disrupt still further their political and social fabric, which was hardly able to digest the enormous sums the oil-exporting states were now earning), at first served to enhance the ruler's autocratic and megalomaniac tendencies. By the end of the century, he kept declaring, Iran would become the world's fourth superpower, thus presumably surpassing in importance and wealth Japan, West Germany, France, etc.! Instead, the last year of his reign found Iran's whole economic and political structure foundering and the emperor stricken by a mortal illness and incapable of deciding on a consistent course of action in dealing with the rising politico-religious ferment. The once imperious King of Kings now looked helplessly for aid and suggestions from the Western capitals that only a short time before he had been treating so haughtily. For its own part, Washington had matched his indecision: the Iranian government was intermittently advised to stand firm and, conversely, to appease the opposition. Apart from everything else, the crash of the imperial regime had thus been a lesson to the whole Third World, whose mostly autocratic and oligarchical governments could not but draw a moral from the Soviet behavior in, say, Ethiopia compared to America's vis-à-vis Iran. The Soviets stuck with their protégés through thick and thin, no matter how unappetizing the character of their rule or how unpopular it was with their own people. In return Russia's friends were expected to behave in an appropriate fashion or, as the case of Afghanistan would soon demonstrate, face unpleasant consequences. By contrast Washington had to be seen as incapable both of disciplining its allies and of helping them in their hour of need. As long as the Shah appeared to prosper, the US government, despite all its misgivings, proved unable to curb his autocratic behavior or to resist his pleas for vast quantities of the most sophisticated American weapons.

This arsenal, quite apart from the strain it put even on Iran's oil-swollen revenues, was quite excessive and superfluous for a country in Iran's position, and of course it would prove of absolutely no use in averting the regime's fall. It was only when the imperial government began to totter that official Washington, as well as the American media, grew increasingly and publicly censorious about its record on human rights, thus contributing to the hapless monarch's indecision and discrediting him still further before his own people. Finally, it was largely because of Washington's advice that the Shah felt compelled to leave Iran. A special American emissary, General Robert Huyser, was dispatched to Iran in January 1979 to see if something could be saved from the wreckage. "Expected at once to organize and prevent a coup d'etat, encourage and restrain the Iranian generals, support Bakhtiar [15] and a military organization that had pledged its loyalty to the shah and to master the subtleties of the Iranian whirlwind, Huyser was out of his element." [16] It fell to General Huyser to advise the Iranian military on Washington's behalf not to resist the Khomeini takeover, their reward for this self-restraint being a wholesale massacre of the upper echelons of the armed forces. Still hopeful of appeasing the new masters, the US government also refused to grant asylum to the Shah. The worst was yet to come, but as of spring 1979 America's current and prospective allies could contemplate yet another example of how this country treats its friends once they fall upon hard times; despite the vastly different circumstances, Washington's unheroic attitude towards the fallen monarch brought poignant memories of its previous dealings with Chiang Kai-shek, Diem, General Thieu . . .

A combination of theocracy and mob rule was now paramount in Iran. In March the country was proclaimed an Islamic republic. But the subsequently erected formal structure of government—president, council of ministers, etc.—would remain but a superficial veneer with real power exercised (in

15. The Shah-appointed prime minister.
16. Michael Ledeen and William Lewis, *Debacle: The American Failure in Iran* (New York, 1982), p. 180.

a chaotic fashion, to be sure) by the Revolutionary Council, Khomeini's puppet. He himself remained the ultimate authority; his pronouncements on the most varied subjects—ranging from the proper apparel for women to the need of strict censorship of the press—assuming the force of law. As a minister stated, "The decision of Imam Khomeini is above that of myself and the government." [17] In addition to his formal (so to speak) exercise of power, a word from the Ayatollah conveyed in sermons by the network of mullahs would launch the city mobs on a rampage, whether against the American Embassy (which already in February had been briefly occupied), some political activists currently obnoxious to Khomeini, or occasionally even members of what passed for the government. The usual vocabulary of politics breaks down when dealing with post–February 1979 Iran. The country was governed not so much by what might be called policies as by waves of emotion generated by religious fanaticism, xenophobia, and anti-Western obscurantism.

As already noted, the Soviets' great concern was that the initial anti-American thrust of the Revolution would yield to a more accommodating attitude towards the US. There seemed little reason for the Kremlin to worry. Washington's ties with Iran had been shattered. On practically every political issue the Revolution took a position diametrically opposed to US interests. The new Iran would no longer support the pro-Western regimes in the Persian Gulf—quite the opposite. Instead of the friendly attitude toward Israel maintained by the Shah, Khomeini endorsed the position of the most extreme elements within the PLO. Thousands of Americans, ranging from members of the military advisory missions to schoolteachers and businessmen, had been expelled or fled the country. The US electronic listening posts on Iranian soil, so important for monitoring the Soviets' strategic arms tests and installations in Central Asia, had been destroyed or seized.

Despite all such discouraging developments, Washington persisted in its hopes of propitiating the Revolution. The few diplomatic and military American officials still left in Iran were

17. Quoted in *ibid.*, p. 197.

encouraged to maintain contacts with official government circles, even though the Ayatollah and his entourage would not meet any representatives of the "great Satan." However, the Iranian army fighting the Kurdish rebels soon developed the need for spare parts for the country's US-made military equipment. The "satanic" character of the US notwithstanding, Teheran's defense ministry turned to Washington for help, and amazingly enough, it was not rebuffed. Some military materiel was ferried to Iran in the summer, and as late as October 1979 a State Department official traveled to Teheran, his mission allegedly to offer it further American help.[18]

The constant stream of anti-American propaganda pumped into Iran by the USSR was thus intended to minimize the possibility of the two once closely allied states resuming some of their previous ties. But while such a possibility could not be taken very seriously outside Washington, what began to worry the Soviet Union beginning in spring 1979 was that the anti-US thrust of the Iranian Revolution had as yet failed to result in any *direct* benefits for Moscow. The Kremlin obviously expected Iran's new masters to turn to it for help in their precarious economic and political situation, yet no such request seemed to be in the offing. Writing in *Pravda*, a Soviet specialist on the Middle East painted an alluring picture of benefits Iran could derive through collaboration with its northern neighbor. The USSR had accumulated much experience in irrigating its own arid areas, experience that could be of great help in turning Iran's semi-desert regions into economic assets. There were rivers flowing through the territories of both countries that could be utilized to produce power, with Soviet technology and experts assisting the Iranians in erecting a network of hydroelectric stations. Finally, "it would be of mutual benefit to unite the energy-producing systems of the two countries." In brief, Iran was being invited to link its economy with that of the USSR and thus to become dependent on the latter to a much greater extent than it had ever been on the US. With the time when it would face diminishing yields from its own fields approaching, the Soviet Union obviously had

18. *Ibid.*, pp. 224–28.

very special reasons for seeking to "merge" its energy resources with those of Iran.

As if mindful of such implications, our writer devoted most of his article to enumerating all the good deeds the USSR had already done for the Iranian people. He rebuffed indignantly the charge recently proffered by CIA director Stanfield Turner that the Soviets were interfering in Iran's internal affairs: "The USSR sided resolutely with the Iranian Revolution, and has made a great contribution to preventing foreign interference in the country's domestic affairs. As for itself, the Soviet Union, in view of its peace-loving policy, never has engaged, nor, needless to say, would engage in such interference. And now with the Revolution having triumphed, the USSR is ready to do its utmost to strengthen still further and expand Soviet-Iranian relations." [19]

Possibly with the example of Khaddafi's Libya before their eyes, the Soviets hoped that even if it adhered to Moslem fundamentalism, Iran's stridently anti-Western regime would find it imperative to seek closer ties with Moscow. But unlike Libya, Iran has had a long, and—despite what our writer would have us believe—far from happy history of dealings with its Russian neighbors. And for all the irrational elements in their makeup, Khomeini and his people were acutely aware that the pro-Soviet and Communist elements in Iran had had their own reasons for joining the anti-Shah coalition, as well as some very specific ideas about the country's future. What those ideas were could be easily imagined on the basis of current developments in Afghanistan. The aged Ayatollah was not mollified by flattering messages from Brezhnev, who, disregarding diplomatic and constitutional protocol, kept addressing him as "the Leader of the Iranian people." The left and the right were one in their detestation of America. But intermittently the Mujahidin ("Fighters for the Faith"), Khomeini's pretorian guards, cracked down on the left. "On August 20 [1979] Tudeh headquarters . . . were occupied by the revolutionary guards, the Communist paper *Mardom* was shut down, and the Communists labelled 'sons of Satan, atheists, the evil of

19. P. Demchenko, "The Horizons of Cooperation," *Pravda*, April 6, 1979.

the earth.' "[20] For the moment, Moscow's entreaties were being scorned.

In Iran, then, the Soviets waited upon the events in the not unreasonable expectation that both internal developments and the country's international situation were bound eventually to turn the tide in their favor. The more pressing problem the Kremlin currently faced was the preservation of the outposts it had already established in the Middle East. By the spring of 1979 some within Soviet officialdom must have already regretted the extent of the Russian involvement in Afghanistan, if not indeed the very coup that had brought that country within the Communist bloc. The proud people who in the nineteenth century had managed to preserve their independence against threats from the world's two greatest empires now showed no signs of acquiescing to what the majority of the Afghans considered as alien rule. In its doctrinaire zeal the Communist government tried to ride roughshod over the country's religious and national traditions, the result being widespread insurrections and guerrilla activities. Large districts fell to the rebels, and even many of those under Kabul's nominal control became "liberated" after nightfall. The government's precarious position was being compounded by factional and personal feuds within its ranks, each party trying to blame the regime's predicament on its rivals and seeking Moscow's support in trying to get rid of them.

For once the Kremlin could reflect that not every American discomfiture must automatically benefit the Soviet Union. The recent developments in Iran had undoubtedly strengthened the determination of the orthodox Moslems in Afghanistan to try to overthrow their government. And even though most Afghans belonged to the Sunni branch of Islam, rather than his own Shiite branch, Khomeini was prompt to embrace the rebels' cause and express publicly his hope that just as the people of Iran had overthrown the puppets of one infidel imperialist power—the US—so the people of Afghanistan would do the same to those of the USSR. Afghan refugees were granted the facilities of Teheran radio to broadcast to their country; a typ-

20. Ledeen and Lewis, *op. cit.*, p. 206.

ical message proclaimed that as "Iran managed to put the de-
mon of Western imperialism to flight, so our nation is struggling against an aggressive army of the criminal Soviet Union
and will not rest until final victory and establishment of the
Islamic Republic of Afghanistan." [21]

Deeply divided among themselves on a number of political
and other issues and poorly equipped by comparison to the
modern Soviet weapons at the disposal of Kabul's army, the
guerrillas still managed to inflict defeat after defeat upon the
government forces, as the latter became increasingly demoralized and suffered from mass desertions. By the end of the
summer it must have been clear to the Soviets that their Afghan protégés were neither united nor strong enough to pacify
the country, or perhaps even to retain power. Since World War
II the USSR had never permitted a Communist regime it had
fully endorsed and supported to be overthrown by its own
people. Could Moscow now afford to let this happen in Afghanistan?

The experience with Sadat's Egypt had undoubtedly served
to increase the Soviets' conviction that their influence in a
given Third World country could never be considered secure
unless it was backed by their own troops or by those of their
trusted friends (e.g., Cuba). Also considered helpful in this respect, though not nearly as much, was the given regime's definitive commitment to Marxism-Leninism, something that—
though no longer a guarantee of loyalty to the USSR, as witness the sad cases of Yugoslavia and China—was still bound
to make the local dictator or ruling oligarchy less likely to
turn suddenly against Moscow and enter upon a sinful liaison
with the West.

Such considerations were probably important in leading the
Soviets to back the 1978 Kabul coup. One must draw the same
conclusion from the events that took place the same year in
South Yemen. That small but strategically located Arab state
practically fell into the Soviets' lap after the British withdrew
from Aden and in general abdicated their previous role east of

---

21. Quoted in *Keesing's Contemporary Archives for 1979* (London, 1980),
p. 29878.

Suez. Long a docile client of the USSR, the People's Democratic Republic of Yemen exhibited some disquieting symp-
toms in 1978. Its President, Salem Ali, his courage possibly
fortified by Saudi Arabia's money, was incautious enough to
indicate that he wanted to establish closer ties with China, as
well as to reestablish diplomatic relations with the United
States. He also planned to seek a reconciliation with North
Yemen, for some time now within the conservative coalition
of Arab states headed by Saudi Arabia.

What ensued was a tale of intrigue and bloodshed that brings
to mind once more how much politics in this age of guided
missiles and space travel partakes of the atmosphere of criminal violence reminiscent of the Renaissance Borgias, or, if one
prefers, of the Chicago gangland of the 1920s. On June 24 the
President of North Yemen was assassinated by emissaries of
the South Yemen government. The pro-Soviet members of the
latter's government blamed, undoubtedly falsely, President Ali
for the crime and proposed to remove him for this breach of
diplomatic etiquette. In the subsequent fighting the anti-Ali
forces, backed by Cuban troops, easily prevailed over the army
detachments that remained loyal to the President. On June 26
Ali and his advisers were executed. There followed a thorough purge of the imprudent President's followers. South
Yemen's new strongman, Abdul Fattah Ismail, announced two
days after the coup that the country would henceforth be ruled
according to Marxist-Leninist principles, and by a single political body that adopted the name of the Yemen Socialist Party.

The Soviets' motives in supporting—if indeed they did not
mastermind—the coup were transparent enough. Loss of South
Yemen would have deprived them of a crucial base in Arabia
and the Horn of Africa. Among his other sins President Ali
had begun to chafe under the Russian-Cuban presence in his
country and sought to ban the use of Aden as a transit point
for the passage of Soviet military supplies and personnel to
Ethiopia. He had, as a matter of fact, been awaiting a special
US mission; but two days before its scheduled arrival, Ali's
presidency and life came to an abrupt end.

In addition to lying across the main oil-shipping lanes from the Middle East to the West and Japan, South Yemen served as a Soviet bridgehead in the Arabian Peninsula, a constant threat to the politically and socially vulnerable regime of Saudi Arabia. The next step in the enlarging of the bridgehead was obviously the absorption of North Yemen, which—for all the American arms procured for it by Saudi Arabia—was hardly in a position to withstand an invasion by its southern neighbor, whose forces would be assisted by Cuban troops and Soviet officers. Border warfare between the two Yemens flickered on and off during 1978. In February 1979 South Yemen's army launched a major drive against the North.

But the timing of this probe was unpropitious. The recent developments in Iran had alarmed the US about the situation in the whole Persian Gulf–Arabian Peninsula area. President Carter authorized massive shipments of weapons and dispatched American military experts to the threatened country. Saudia Arabia put its army on alert. Even such pro-Soviet regimes as those of Iraq and Syria expressed their opposition to what they saw as an attempt to impose Communist rule on yet another Moslem state. The probe thus threatened to ignite a major international crisis, as well as to alienate the Soviet Union's supporters within the Arab world; for the governments of those countries—e.g., Iraq and Algeria—drew a line between being Moscow's allies and its puppets (both on occasion dealt quite unceremoniously with their own Communists). The Arab League sponsored negotiations between the two Yemens, and a cease-fire between them took effect on March 16. Incongruously enough, these recent enemies then began talks about a peaceful merger of the two countries. Such unification, as had been the case with so many planned and attempted mergers between Arab states, was bound to prove a mirage. But the whole sequence of events strongly suggested that for the moment the USSR had decided that it was more prudent to pursue its goals in the Arabian Peninsula through other than military means.

In some respects, the 1978 drama of South Yemen was a preview of the events that would take place in Afghanistan in the latter part of 1979. On September 17 the Soviet press reported the arrival in Aden on the preceding day of Prime Minister Kosygin. This mission of the Kremlin's number two man to South Yemen, which until a few years before would have hardly rated a visit by a minor Soviet functionary, indicated the importance the little country had assumed in Moscow's foreign policy strategems. On the same day a TASS dispatch from Afghanistan reported that a special plenum of the Central Committee of the People's Democratic party of Afghanistan had acceded to a request by Noor Mohammed Taraki "that in view of the state of his health, he be released from his party and state offices" and had appointed Hafizullah Amin in his place.[22] Taraki's health was indeed in unenviable shape. Two days before the announcement, shots had rung out in Kabul, and Moscow's erstwhile satrap either had been or was about to be executed. For reasons of decorum, it was only on October 10 that the *Kabul Times* announced (on its last page) that "Noor Mohammed Taraki, former president of the Revolutionary Council, died yesterday morning of a serious illness, from which he had been suffering for some time."[23]

In their separate ways, Iran on the one hand and South Yemen and Afghanistan on the other stood as of 1979 as vivid examples of the differences between the superpowers' reactions to threats to their interests and influence in the Third World. Increasingly (and the most drastic demonstration of this would come at the end of the year in Afghanistan), the USSR was resorting to force to protect its position in states that had opted for, or stumbled into, close relations with Moscow. Such had not been the case, say, in the 1960s, when some regimes— e.g., Ghana's and Guinea's, which at one time had leaned heavily on the USSR—had sought and achieved a more balanced position between East and West. To this one may contrast what happened in South Yemen when its President attempted to reduce his country's dependence on the Soviet Union.

The US, on the contrary, proved unable to employ its still

22. *Pravda*, September 17, 1979.

23. Anthony Arnold, *Afghanistan: The Soviet Invasion in Perspective* (Stanford, 1981), p. 85.

enormous military and economic power to stop or moderate the Iranian Revolution. The contrast was all the more instructive because in itself South Yemen was of little importance on the chessboard of world politics, while Iran was crucial to the balance of power in the Middle East. The Arab country could not affect significantly the security or economy of the USSR; while the curtailment of Iran's oil production led to a shortage and rocketing prices of that vital commodity in the entire non-Communist world, thus dealing the economy of America and its allies a blow at least as severe as that rendered by OPEC's 1973–74 exactions.

Even when Washington did bestir itself to react to a perceived threat to America's interests, its actions were often belated and of limited effectiveness. It simply could not match the dispatch and decisiveness exhibited in analogous cases by Moscow. When the Carter Administration invoked emergency powers to announce that it was sending arms and American military advisers to North Yemen to defend it against the South, it ran into considerable criticism in Congress, and the attendant publicity proved embarrassing to both the US and North Yemen. Covert operations—i.e., by the CIA in such cases—were now illegal unless Congress had been notified, which meant in practice that they could not remain secret for very long and would become a subject of heated debate in this country and of indignant denunciations by Moscow. Nobody could tell for sure how many Cuban soldiers and Soviet and East German military officers were in South Yemen. But to propose even a miniscule American military presence in a threatened area—in this case seventy to a hundred military experts in North Yemen—invariably gave rise to fears and charges that "another Vietnam" was in the offing. Perhaps not surprisingly, once the immediate crisis had passed North Yemen found it judicious to hedge its bets and to request arms from the USSR as well.

To summarize: it had become much easier and hence tempting for the Soviet Union to act forcefully to expand its spheres of interest, and more difficult for the US to act in a like manner to protect its own. Had they had a sense of humor, the

men in the Kremlin may well have congratulated themselves
on their ability to do to the West what the Americans had
once tried to do to them: employ the strategies of containment
and "rollback." With an occasional and for the moment unavoidable exception (i.e., Egypt), any challenge to the Soviets'
enclaves of influence and power throughout the world, whether
coming from local conditions or promoted by US diplomacy,
was being decisively rebuffed. At the same time the ghost of
John Foster Dulles must have watched unhappily as *American*
influence was being rolled back, especially in the Third World.
The Soviets had developed a very effective technique of nib-
bling at the edges of what not so long before had been undis-
puted spheres of interest of the West: southern Africa, Ethiopia, Arabia.

The USSR was also an indirect beneficiary of the fairly general (in the non-Communist world) weakening of the traditional forms of political authority and culture, with the attendant exacerbation of radical, ethnic, and religious conflicts,
international terrorism (often feeding upon the former), and
growing skepticism about the value of democratic and liberal
institutions.

With such ailments afflicting the West, there is little wonder
that many countries in other parts of the world lacking a long
experience of representative institutions would experience social and political upheavals that more often than not brought
into power authoritarian regimes. If of the left variety, a regime would immediately become a subject of concern to the
US and of interest to the USSR; if of the right, it usually expected American understanding and support as a reward for
not being pro-Soviet. Even in the heyday of the Cold War, large
segments of public opinion in the West found it distasteful
that the competition with the Soviet Union and Communism
should make the US an ally of governments such as Franco's
in Spain and Chiang's in Taiwan. In the wake of Vietnam,
there came even more insistent and vociferous demands that
the US ought not to aid countries whose rulers could not display irreproachable democratic credentials, nor should it con-
nive, whether openly or covertly, with military or right-wing

elements seeking an overthrow of their leftist, pro-Soviet regimes. Washington was widely blamed for the help it rendered to the opposition forces in Chile prior to the coup that
toppled Allende's government in that country. Carter inaugurated his administration with a firm resolve to be eqully censorious about both Communist and non-Communist violators
of human rights, but he soon had to face the fact that while
his reprimands on this count had little effect on the former,
they tended to weaken and/or alienate the latter. A classical
demonstration of the difficulties of practicing absolute virtue
under the conditions of twentieth-century politics is the case
of Argentina. Its military rulers bridled at Washington's quite
justifiable criticisms of their internal policies, and they felt no
compunctions about greatly increasing Argentina's wheat sales
to the USSR, thus helping to offset the effect of the American
grain embargo imposed in retaliation for the Soviet invasion
of Afghanistan.

But whatever the American discomfitures and Soviet successes on the world scene, the Kremlin was far from down-
grading the importance of direct US-USSR relations, or, to put
it differently, of preserving the appearance, if not the substance, of detente. On the contrary, the expansionist thrust of
its policies in the 1970s made it more urgent, in Moscow's
view, to minimize the chances of a confrontation between the
two powers. Reassured about Soviet intentions concerning the
West, relieved of the nightmare of a nuclear clash, America
was more likely to acquiesce in Russian gains elsewhere. Apart
from its inherent importance, a strategic arms agreement would
serve as a tranquilizer, inhibiting the democracies' anxieties
about the changing international scene, as well as undercut-
ting the arguments of those who called for a "tough" anti-
Soviet line. Whatever some wretched Kremlinologist or even
government official might try to tell him, the average American or West German voter could not become overly excited
about what was happening in Angola or Afghanistan; his main
concern was his own country's security and the danger of a
nuclear holocaust, and he hoped that East-West differences
could be settled through a dialogue and at the conference ta-

ble. So even the mere picture of the leaders of both sides sitting down together was bound to have a calming effect on the international scene. And since introspection is not a strong point of the Soviet leaders, it is only reasonable to suppose that they themselves saw no incompatibility between their overall policies and the desire to avoid pushing the world to the brink.

The craving for a nuclear agreement and for patching up detente was not absent on the other side. "Carter had said from time to time that 'if only I could get my hands' on Brezhnev, it would surely do some good. The President had for a long time nurtured the conviction that much of the bad blood between Washington and Moscow could be cured by personal diplomacy. He had been confident that the seemingly insurmountable obstacles of SALT would disappear, if only he and Brezhnev could sit down and reason together, look each other in the eye and talk man to man about the problems of war and peace."[24] By the time Carter and Brezhnev met in Vienna on June 15, 1979, the "insurmountable obstacles of SALT" had in fact been overcome, and the new treaty on the limitation of strategic arms was ready for their signatures.

Though the technical negotiations leading to SALT II were theoretically conducted in secrecy, the details of the ill-fated agreement had been known for some time and had already aroused a tempestuous debate in the United States. In its essence SALT II followed the guidelines formulated in Vladivostok as far back as 1974, though the contentious issues of the Soviet Backfire bomber and the cruise missile, then laid aside, were now dealt with, not in the treaty proper, but in additional documents attached to it.

In its substance, SALT II differed from its 1972 predecessor in that it included qualitative as well as quantitative limitations on nuclear arms. Soviet intransigence paid off insofar as the US acquiesced in the USSR's retaining over three hundred "heavy" ICBMs against none for the US. When it came to the other "light" (though in most cases still with vastly more de-

24. Strobe Talbot, *Endgame: The Inside Story of SALT II* (New York, 1979), p. 10.

structive power than the Hiroshima bomb) missiles, each side was allowed to test and deploy one more system beyond those it already possessed as of May 1, 1979. This provision presumably opened the way for the US to offset the Soviet Union's alleged first-strike capabilities by developing the MX missile. Soviet obduracy also prevailed on the issue of cruise missiles; the protocol to the treaty limited such projectiles, if sea- or land-launched, to a range of 600 kilometers (360 miles), with only those mounted on bombers allowed a longer operating radius. As for the total number of strategic nuclear arms launchers of all kinds—land, undersea, and airborne (i.e., bombers)—neither side was to exceed 2400, this ceiling to be reduced by January 1, 1981, to 2250. Of these, not more than 1320 were to be MIRVed.

The thorny problem of the Backfire bomber was handled in a rather unusual fashion. Brezhnev handed Carter a written statement declaring that the plane did not have an intercontinental range, that the USSR would refrain from modifying it to endow it with such, and that its production rate would not be increased above thirty per year. The American critics of SALT, the Soviets knew, had made a great fuss over the bomber, and Brezhnev probably felt that he was going beyond the call of duty in offering this assurance and revealing data about a weapon that was not strategic, and hence should not have been on the agenda at all. There remained a perplexing (hopefully never to be answered) question: should a nuclear conflict erupt, would the Soviet command remember Brezhnev's pledge and refrain from using the bomber, which from a Siberian base could easily reach the western US and then come back? But the letter was intended to make it easier for Carter to get SALT II accepted by the Senate (though in fact it did not).

There could be relief on the American side that despite their previous arguments (e.g., during Vance's 1977 visit to Moscow), during the latter phases of negotiations the Russians abandoned their demand that the "forward-based systems" (i.e., the tactical nuclear weapons under NATO) be included in the SALT calculations. To anyone familiar with the Soviets' ne-

gotiating techniques, this should have been a tip that Moscow
had something up its sleeve concerning medium- and
intermediate-range nuclear weapons, and such indeed, it later
became clear, was the case.

Beyond signing SALT II, the four-day Vienna meeting af-
forded the leaders of both countries an opportunity to review
the whole spectrum of world politics. The two delegations appeared well suited to high-level talks on a variety of sensitive
subjects. In addition to a crowd of technical experts, Brezhnev
brought with him three Politburo members—Gromyko, Con-
stantine Chernenko, currently an ascending star in the Soviet
political firmament, and minister of defense Dmitri Ustinov—
as well as the Soviet chief of staff, Marshal Nicholas Ogarkov.
Carter was accompanied by Vance, Brzezinski, and Marshal
Ogarkov's American counterpart, Gen. David Jones. Yet for all
this assembly of luminaries, the political side of the discussion did not advance much beyond the proverbial exchange
of views, with both sides recording their complaints and misgivings concerning each other's position on the Middle East,
Southeast Asia, the Cubans in Africa, etc. The American President's position was in the Soviets' eyes quite different from
that of his predecessor at the 1972 summit in Moscow. Nixon
had then arrived after his dramatic breakthrough on China,
and his political position at home was believed to be well-
nigh unshakeable. Vienna took place after a series of foreign
policy setbacks for the US, relieved only by Carter's recent
success in persuading Egypt and Israel to consummate the
Camp David agreement and to sign a peace treaty. The President's chances for reelection appeared at best problematical—
his foreign policies were under scathing attack, recession and
inflation were assuming ever more serious proportions. The
"China card" had been bid, but up to now with meager results. Much as they wished that SALT II should be ratified,
Brezhnev and Co. were in no mood to present the President
with some spectacular offering that he could then display at
home to convince the Senate and public opinion that detente
was alive and well. The experience of the last six years strongly
suggested that the Americans responded (to be sure within

certain limits) to pressure and were impressed by Soviet in-transigence—witness their dropping the whole business about the Soviets' "heavy" missiles and their acquiescence in the Cuban presence in Africa. Offer them some substantial concessions, and they would convince themselves that their Chinese gambit was working and start pressing you harder on human rights or God knew what else.

The Russians also sensed that American foreign policy lacked that unity of leadership and sense of direction that, whatever else one can say about it, had characterized the Nixon-Kissinger era. Then, and until the disasters of Watergate and the OPEC price rise threw both the leadership and the economy of the West into turmoil, the Soviets had tended to be more mindful of US sensitivities and potential reactions and hence more circumspect in their own policies.

Now they acted as if they did not believe there was very much that the Carter Administration could do for, or, conversely, to the USSR. Even with respect to the treaty, it was obvious that the President could only propose, and it was the Senate that would dispose. The American officials who like the army of reporters may have expected some momentous statement—a breakthrough in East-West relations—to emerge from the summit could derive but little comfort from the tepid and commonplace tone of the joint communiqué issued at its conclusion.

That perennial hope of American politicians that personal contacts at the highest level help bridge the gap and melt animosities between the two sides also failed to come true. Unlike Khrushchev at that other Vienna summit, the Soviet leader did not try to bully his American opposite number. But in his formal speech in the American Embassy on June 16, Brezhnev served plain notice that he and his colleagues did not propose to modify Moscow's current policies and that the conference must stick to the business at hand, i.e., SALT. "Attempts also continue to portray social processes taking place in one country or another or the struggle of peoples for independence and social progress as 'Moscow's intrigues and plots.' Naturally Soviet people are in solidarity with the liberation struggle of the peoples. . . . We believe that every people has a right to decide its own destiny. Why then pin on the Soviet Union the responsibility for the objective course of history, and what is more, use this as pretext for worsening our relations?"[25] When it came to substantive discussions of international issues, the Soviets thus simply replayed old records.

Slow and ponderous in his movements, Brezhnev seemed to symbolize the ossification of the Soviet political system. But, by the same token, he and his not much younger or more nimble chief aides exuded a feeling of sternness and fixity of purpose. American observers looked eagerly for any signs that the Russians might be mellowing. But even when posing for pictures with their US counterparts, the Soviet leaders eschewed any hint of conviviality, obviously trying to convey the impression that this was serious business and that no amount of sociability could make the USSR swerve from its course. This obviously prearranged show of solemnity did have its comic undertones. "God will not forgive us if we fail," said Brezhnev to Carter. Such references to the Deity—though in Russian they are but a turn of phrase rather than an indication of religious feeling—have always intrigued Americans. But with Carter being a born-again Christian, Brezhnev's aides must have feared that the statement might be misinterpreted or even provide an opening for the President to bring God into the discussion of human rights or Ethiopia. Alertly, Gromyko and his press spokesman tried to laugh off and explain away the remark.

As if to confound the notion of Western decadence, so often expatiated upon in Brezhnev's speeches (the ones for domestic consumption), the American leader ran around the American Embassy early in the morning of the first day of the meeting. Thereafter he refrained from jogging, possibly realizing that such demonstrative displays of physical vigor might not be appreciated by his Soviet counterpart, almost twenty years older. But in a sense, this was a suitable retribution for the Brezhnev of 1972, then much more fit, taking Nixon and Kis-

25. Quoted in *Department of State Bulletin*, no. 2028 (July 1979), p. 51.

singer for that hair-raising spin in his automobile. And Carter did get his hands on Brezhnev: after the signing ceremony the two men embraced and kissed—the first time an American President had indulged in this quaint (to the Anglo-Saxon eye) Slavic custom.

Moscow's eschewing of any show of excessive cordiality was also designed to impress upon the Americans that it would not tolerate any tampering with the treaty's provisions by the Senate. Whether in fact this was actually so is open to doubt. Superb bargainers that they are, the Soviets often cling tenaciously to their original negotiating position, only to make a last-minute concession or plea for a compromise, usually on an issue not of overwhelming importance to themselves, so as to smooth the passage of the rest of the agreement. SALT II did contain such a "sleeper," insofar as the Backfire bomber was *not*, as we have seen, counted as part of the Soviets' strategic launchers quota. Everyone knew that this was *the* provision on which the anti-SALT forces in the Senate were about to concentrate their fire. It would not then have been too surprising if, had the Senate insisted on an amendment to the treaty classifying the Soviet bomber as a strategic weapon, Moscow, after exhibiting suitable indignation, conceded the point. Compared with land- and submarine-launched missiles, the nuclear armed bomber is a clumsy strategic weapon, both in the time it takes to reach its target (several hours for an intercontinental mission, as against thirty minutes or less) and in terms of its vulnerability to ground and air defenses. Thus it would not have hurt the USSR too much to discontinue production of the Backfire, or offer to trade it off against an appropriate American quid pro quo. Such a step would then presumably disarm all but the diehard Senate opponents of SALT II.

For the moment the Soviets waited and watched the unfolding of the administration's campaign to "sell" the SALT package to American public opinion and to secure those sixty-seven Senate votes necessary for ratification.

What stood in the way of ready acceptance of SALT II by the American public and Congress? First of all—and this eventually would prove decisive in dooming the agreement—there was the widespread feeling of frustration over the whole course of Soviet-American relations since the beginning of detente in 1972. It is safe to assume that had there been no Cuban troops in Angola and Ethiopia, no Vietnamese grab of Cambodia, and (the final blow) no Soviet invasion of Afghanistan, whatever objections there might have been to the actual provisions of the treaty, it would have passed into law, probably with some modifications as indicated above. The Carter Administration tried valiantly to separate the overall issue of Soviet policies from that of the treaty, which it proclaimed to be in the national interest regardless of what Russia was or was not doing in Africa and Asia. But for many Americans, the setbacks, disappointments, and humiliations experienced by the US throughout the world ever since Nixon and Brezhnev had concluded their compact in 1972 were seen as the result not, as Brezhnev would have had it, of "the objective course of history," but of deliberate Soviet policies. You would not be punishing the USSR by rejecting the treaty, nor, conversely, conferring a boon upon it by accepting it, pleaded its proponents; it was a sensible step in the direction of minimizing the chances of a nuclear war, and entirely compatible with US security. But whatever the logic of such pleading, eventually it was to prove unavailing.

The second large area of American criticism of SALT II was centered on the allegation that its provisions would render this country vulnerable to a Soviet preemptive strike, one that under the quotas established in the Vienna agreement could not be reliably deterred by the weapons at the disposal of the United States. This argument in turn partook of two elements: one that could be described as philosophico-political, the other based on deductions from statistical and technical data.

The former rejected the notion that the rulers of the USSR abhorred the prospect of a nuclear war as strongly as did the people and governments of the West. In fact, it was argued, the Soviet Union was preparing for such a war, and its civilian and military leaders believed that it could be won in a meaningful sense of the term. In support of this thesis, its proponents cited the character and extent of the Soviet civil de-

fense preparations. Unlike the US, where any notion of minimizing the consequences of a potential nuclear attack was
derided or evoked an anguished reaction, the USSR was going
ahead with elaborate measures designed to deal with such an
eventuality: a wide network of underground shelters, disper-
sal of industrial plants, civil defense drills, etc. To be sure the
Kremlin was aware that in an all-out war, no defense could
save the USSR from vast devastation. But then, the argument
continued, the Soviet rulers' idea of what constituted "unacceptable" losses was quite different from what was believed
on this subject in America. World War II had cost the Soviet
Union over twenty million lives, yet the country emerged victorious, the Communist regime stronger than ever. And so the
prospect of losing twenty to thirty million people in a US retaliatory attack after the USSR had struck first was viewed by
Moscow as something it could afford; the Kremlin expected
not only to survive, but to emerge the world's master. "Experience suggests that as of today, the USSR could absorb the
loss of 30 million of its people and be no worse off than it had
been at the close of World War II. . . . All of the USSR's cities
with multimillion population could be destroyed . . . and
provided that its essential cadres had been saved, it would
emerge less hurt in terms of casualties than it was in 1945." [26]

The statistical-technical part of the indictment of SALT II
sought to add a note of urgency to the above picture. Thus it
was claimed that despite numerical parity in the strategic
launchers, the treaty would increase, potentially catastrophically, this country's vulnerability to a nuclear attack by the
USSR. Given the Soviets' rapid progress in MIRVing and improving the accuracy of their ICBMs, as well as their greatly
superior megatonnage, a situation was bound to arise in the
1980s—the famous "window of vulnerability"—when the Soviet Union would acquire the capability, with just part of its
land-based missiles, to wipe out *all* of the US ICBMs.

But surely, an advocate of SALT II retorted, only utter in-
sanity could prompt such a Soviet surgical strike, which would

26. Richard Pipes, "Why the Soviet Union Thinks It Could Fight and Win
a Nuclear War," *Commentary*, July 1, 1977, p. 32.

still leave the US with a vast quantity of nuclear weapons and expose the Soviet Union to American SLBMs (submarine-launched missiles) and heavy bombers wreaking death and devastation on its cities. Hardly, continued our critics: was a president of the US likely to order a retaliatory attack, knowing that at the most it would kill "only" some 30 million people, leaving the Soviet military and political machine able to function, while the USSR still had enough missiles literally to destroy the US, bring death to upward of 100 million Americans (this country being much more heavily urbanized than the Soviet Union), with lethal radiation and epidemics completing the holocaust?

Few of those evoking such grisly visions intended to have them taken literally, and if pressed they would allow that their Armageddon scenario contained some unproven (and unprovable) assumptions, as well as inherent improbabilities. Still, their argument went on, the mere fact that this scenario could be sketched meant that SALT II in its present shape was bound to have a profoundly destabilizing effect: it would embolden the USSR to pursue even more aggressive policies and tend to inhibit the US from standing up to Moscow in any future confrontation.

The whole debate must have been watched by the Kremlin with mixed feelings. It was not entirely undesirable that some Americans should portray the USSR as a nuclear colossus capable of destroying the United States at some future date, while itself remaining relatively invulnerable to retaliation. In the heat of the debate the Americans tended to forget that they were talking not only to each other; Western Europe and the rest of the world were also listening, and the conclusion some of those countries might draw was not that the US must become stronger, but that they should be more accommodating to the Soviets' wishes.

At the same time the USSR had very strong reasons to wish that the strategic arms agreement would come into force, though not for reasons being attributed to it by the SALT opponents on this side of the ocean. In the late 1940s Winston Churchill observed that the Soviet Union did not want war, it sought the fruits of war. And so in the late 1970s the USSR sought the fruits of its nuclear might, but no one conversant with the Kremlin's way of thinking and operating should have credited it with eagerness to plunge into the frightful uncertainty of a nuclear war. Were Brezhnev and his colleagues to speak frankly when asked why they were seeking a margin of nuclear superiority and honing up their civil defense if they were not planning to start a war, they would have answered that the reasons were obvious: it was of great political and psychological importance to the regime, both internally and in foreign policy, to feign confidence that it could survive and prevail in a nuclear conflict. The other part of the Soviets' candid answer would have astounded the average American: granted that the US was not likely to attack first, it was still preferable to have nuclear superiority exercised by a power whose political and military decisions were made by a handful of prudent and experienced people, acting upon rational criteria, rather than by a disorderly and volatile democracy, prone to sudden gusts of political passion that might drive its government to actions both irrational and incalculable in their consequences.

In Moscow's view, there were other solid advantages to be derived from SALT II: it offered fairly firm guarantees against some American breakthrough in technology that would enable the US to recoup its edge in nuclear weapons; and it promised to breathe new life into detente, with all the consequent economic benefits for the Soviet bloc.

Finally, there was another and very major reason for seeking an agreement with the US, but one about which the Soviets felt they could not afford to be explicit: China. The case for Soviet superiority over the US in certain categories of nuclear weapons could have been made more convincing had the Kremlin avowed the full depth of its concern over China's nuclear progress and prospects. Some idea as to how pervasive this concern was can be gathered from Soviet technical literature. "Beijing's leaders, who spend 40 percent of China's budget for military purposes, are hastily developing its nuclear, as well as conventional, arms potential. China, which up to the middle of 1978 has carried out twenty-three nuclear tests in the atmosphere, has accumulated . . . several hundred nuclear warheads (atomic as well as hydrogen), installed forty intermediate-ballistic-missile launchers (average range, 3000 kilometers), has tested intercontinental missiles (13,000 kilometers in range), has launched its first diesel-powered submarine, equipped with ballistic rocket tubes, and is developing ballistic missiles powered with solid fuel. . . . The Beijing leaders expect to turn China into a superpower by the end of the twentieth century and to do so primarily in order to expand and modernize the country's military potential with the active help of the imperialist powers. The military circles of the US, on their part, have shown readiness to accommodate Beijing, and especially to help it with military technology (whether directly, or through the NATO allies and Japan)."[27]

While the Americans discussed the possibility of this country becoming especially vulnerable to a nuclear strike sometime in the 1980s, the Soviets may have ruefully reflected that in the 1990s, their own country could face a similar predicament vis-à-vis China, and indeed that "window," once "opened," would be extremely difficult to close.

China was very much on the minds of the Soviet plenipotentiaries and experts at each stage of the SALT negotiations. The treaty itself had one very oblique reference to this most perplexing (in the Soviet scheme of things) problem. Article 13 stipulated that "each Party undertakes not to assume any international obligations which would conflict with the treaty."[28] That, in conjunction with the 1968 nonproliferation treaty, which prohibited the possessors of nuclear weapons from transmitting such arms or technology to any other powers, would provide the USSR with a strong case for charging the US with violation of its treaty obligations should Wash-

27. Institute for the USA and Canada of the Soviet Academy of Sciences, *Global Strategy of the USA Under the Conditions of the Scientific-Technological Revolution* (Moscow, 1979), p. 75.

28. U.S. Arms Control and Disarmament Agency, *Arms Control and Disarmament Agreements* (Washington, 1980), p. 227.

ington show the slightest sign of assisting China in developing its nuclear arms and delivery systems.

It is also quite likely that SALT II was viewed by Moscow
as a way of reopening in the future a discussion of the perennial problem of "the third power." In Vienna the two governments signed a statement of guidelines for negotiations on
SALT III, which would presumably come into force when
SALT II expired in 1985. The prospectus included in the
agenda "steps . . . to ensure the equality and equal security
of the Parties" as well as the pledge that "the Parties will also
consider further joint measures, as appropriate, to strengthen
international peace and security, and to reduce the risk of out-
break of nuclear war." [29] Beijing already disposed of missiles
capable of hitting important targets in Central Asia and Siberia; by 1985 it would presumably have some that could reach
Moscow and Leningrad. How could the security of the USSR
be adequately gauged by measuring its nuclear forces only
against those of the US, while China's nuclear armaments remained outside any controls? Should not the Soviet Union and
America at some point take appropriate joint measures to deal
with the problem?

Such fears and concerns of the Kremlin were given but little
consideration in the debate about SALT II, which preoccupied
Congress, as well as the American public, throughout the
summer and fall of 1979. To strengthen the case for the treaty,
the administration emphasized its determination to proceed
with the development and production of the MX, the missile
that was supposed to take the curse of vulnerability off American land-based strategic forces. Much more powerful than
their Minuteman predecessors, some 200 MX missiles installed upon mobile ICBM launchers were to circulate in sub-
terranean corridors seeded with a multitude of silos. By removing any presumption on the Soviets' part that they could
in one strike destroy the entire stock of US ICBMs, MX would
restore the credibility of American strategic deterrence.

The details of the proposed new weapons system boggled

29. *Ibid.*, p. 236.

the imagination: it would involve digging up vast areas in several states; its cost was conservatively estimated at $50 billion. But would it, if ever undertaken, prevent that "window of vulnerability" from opening? The opponents of the scheme were quick to point out that under the Vienna agreements, both sides were prohibited from deploying or flight-testing any mobile ICBM launchers prior to January 1982. Thus at best, and assuming that the Pentagon could prevail over the objections of the USSR as well as those voiced by the voters of Utah, Nevada, and Colorado, the proposed system could not become operational before the end of the 1980s.

In retrospect, many of the arguments adduced both for and against the ratification of SALT II must strike us as rhetorical, rather than as addressing the real issue. The partisans of the treaty were clearly oversimplifying the problem when they argued that this country should not be at all concerned about the potential vulnerability of its land-based strategic forces, or that such vulnerability in fact did not exist. On the other hand, those who opposed the treaty were prone to exaggerate when depicting the hideous consequences that might follow upon its ratification. It is worth noting that as of March 1982, the Reagan Administration, whose chief figures opposed the agreement before and during the elections, had neither adopted nor proposed a single measure that contravened the letter or spirit of SALT II, and indeed both sides up to this point chose to abide by its provisions, even though it had not been officially ratified.

Instead of listening to prophets of doomsday on the one hand and advocates of complacency on the other, what should have been of prime importance to the US in examining the treaty was its potential impact on NATO. Moscow's hopes on that score were transparent enough: that the Vienna agreements would weaken still further the Europeans' reliance on the United States, as well as raise fresh doubts about the credibility of the American "nuclear umbrella" over Western Europe. The latter might thus become more accommodating to the Soviets' wishes, expand its trade with and credits for the Soviet bloc, and be more circumspect about doing the same for China.

The Western European capitals were then faced with a dilemma: they welcomed SALT II as reviving detente, but did not wish improvement in US-Soviet relations to be purchased at the price of their own security. The solution seemed to lie in strengthening NATO's theater forces. Such strengthening was felt to be urgent in view of the greatly increased conventional and especially nuclear power of the Warsaw Pact. It now became clear why the Soviets had dropped their original demand, or feint, that theater nuclear weapons be included under SALT II. They had been deploying a powerful intermediate-range ballistic missile, the SS-20, that would provide the Warsaw Pact armies with superiority in nuclear as well as conventional arms over NATO. Belatedly the US and its allies realized that while worrying about the strategic arms picture, they had neglected what was happening in Europe. As a NATO communiqué of December 12, 1979, put it: "During this period, while the Soviet Union has been reinforcing its superiority in long range theater nuclear weapons, both quantitatively and qualitatively, Western capabilities remained static. Indeed these forces are increasing in age and vulnerability and do not include land based long-range theater nuclear missile systems."[30] To make up for the lost time, a special meeting of NATO foreign and defense ministers decided on that same date to proceed with a plan to deploy 108 Pershing II ballistic and 464 ground-launched cruise missiles in Europe. If and when functional (they were supposed to be installed in 1982), they would be the first ground-based NATO nuclear missiles capable of reaching the European part of Russia.

Especially on this last count, the NATO decision was bound to evoke a violent Soviet reaction. And perhaps the word "decision" is not quite appropriate. Some of the allied governments, notably those of the Netherlands and Belgium, were less than enthusiastic about having intermediate missiles, in fact any nuclear weapons, deployed on their territory. The other NATO powers, principally West Germany, did recognize the acute need for offsetting the recent expansion of the War-

30. Quoted in *Department of State Bulletin*, no. 2035 (February, 1980), p. 16.

saw Pact's nuclear armory. If the Russians really felt threatened by the prospect of their cities facing the same danger that their own SS-20s already posed to London, Paris, or Hamburg, well, let them negotiate for the elimination or limitation of tactical nuclear weapons by both NATO and the Warsaw Pact. But no one knowing how tenaciously the Kremlin holds on to an advantage it has once acquired could believe that such negotiations would be easy or expeditious. The Western European public had grown accustomed, almost to the point of being oblivious, to the short-range (battlefield) nuclear weapons already in the hands of NATO. But as already seen in the case of the neutron bomb, any proposed introduction of a new variety of nuclear arms on the Continent in the face of not-so-veiled Soviet threats was bound to reawaken and intensify the Europeans' anxieties and make many question the desirability of having *any* kind of nuclear arms in their midst. By the end of 1979 some Pentagon planners must have been quietly cursing themselves for not seeing through the Soviets' game and taking them up on negotiating about tactical nuclear weapons in conjunction with SALT II. Once again Moscow had stolen a march on the West, and would employ all its diplomatic and propaganda resources to retain or sell dearly its advantage.

It was, however, not so much Soviet arms or the complex pattern of international developments during the latter part of the year that doomed SALT II. Had the events to which we shall now allude not taken place, the Vienna agreements would in all likelihood have been ratified by two-thirds of the Senate, possibly with some amendments that Moscow, after making ferocious grimaces and noises, would have accepted. The Kremlin placed great importance on the ratification of the treaty, and during its discussion by the Senate tried to observe what was by its lights considerable restraint in its international posture. It is probably not unconnected with the SALT debate that in the course of 1979 more than 50,000 Jews, the highest number ever, were allowed to emigrate from the USSR.

How issues having nothing to do with the substance of the treaty could affect its fortunes became evident by late summer, during the brief imbroglio over the presence of Soviet troops in Cuba. With the American political scene already heating up as the 1980 presidential elections approached, administration opponents were about to publicize the fact that the USSR had a combat unit stationed on the island. It was therefore considered advisable to have the announcement come first from a pro-administration source, to rebut in advance any insinuation that once again Washington was asleep while the Soviets stealthily penetrated the Western Hemisphere. There followed a flurry of excitement, with President Carter and Secretary Vance issuing stern warnings that the presence of the "Soviet brigade" on this side of the ocean was "not acceptable," that this intrusion might jeopardize the fate of SALT II, etc. For its part, Moscow retorted that the "brigade" was actually a training unit, that it had been there all along, and that the whole business was being blown all out of proportion. On October 1, the President broadcast a confident message to the nation: he had been reassured by the Soviet government that the troops in question were in fact a training unit and that they did not pose any threat to the security of the republic. And the US would remain vigilant so as not to allow any hanky-panky in the Caribbean. In retrospect, it is difficult to believe that the presence of the Soviet contingent, some 2000–3000 soldiers, had not been known to the US government before the whole fuss originated. As to the reason for their being there, the most likely explanation is that they provided additional insurance against anything untoward happening to Castro and his regime.

With this Cuban mini-crisis (as compared to 1962) over, SALT II seemed set to continue its tortuous progress through the Senate. But by now, anything that struck at America's already brittle self-confidence in foreign affairs, whether it had anything to do with the USSR or not, was bound to affect US-Soviet relations. On November 4 a crowd described as being composed of students seized the American Embassy in Tehran, imprisoning fifty-three US citizens found inside it. This act, perpetrated allegedly in retribution for the recent admission of the ex-Shah to the United States for medical treatment, was immediately approved by Ayatollah Khomeini. The leader of the Iranian Revolution simultaneously fired the prime minister, Mehdi Bazargan, whose transgressions had consisted of trying to steer a moderate course in relations with the US and who at a recent ceremony in Algiers had been photographed shaking hands with the security assistant, Brzezinski.

There are grounds to believe that the seizure of the embassy had been planned some weeks before the Shah's admission to the United States.[31] Anti-Americanism was one common ground on which the diverse elements that coalesced to carry out the Iranian Revolution—the leftists and Moslem fundamentalists—could unite, and the lingering humiliation inflicted upon the "Great Satan" thus became a political asset for the chaos that passed for the government as it struggled against secessionist movements, mounting economic distress, and, later on, aggression by Iraq.

Strictly speaking the seizure—unless promptly repudiated and the hostages released and indemnified—was an act of war, and up until a few years previously a major power would normally have responded to it with an ultimatum and, if this was rejected, with military measures. But on November 6 the White House announced that the US had ruled out the use of force, and on the eighth Secretary Vance pronounced what had by then become a US liturgical formula when confronted with a sudden crisis: "It is a time not for rhetoric, but for quiet, careful, and firm diplomacy."[32] Well, the next fifteen months were to witness a lot of American rhetoric, but hardly the kind of diplomacy that Vance had promised. The nation kept an anxious vigil over the hostages as Washington tried intermittently to placate and to threaten Teheran, kept appealing to world public opinion, and at one point staged an abortive rescue mission. No one wanted to endanger still further the unfortunate prisoners' lives, but few could overlook the disastrous consequences of the affair to American prestige, as well as to what had previously been the accepted norms of international life.

31. Ledeen and Lewis, op. cit., p. 229.
32. Ibid., p. 230.

For our purposes it is important to register a strong presumption that the whole affair weakened the Soviets' resolve to observe restraint in their foreign activities while SALT II was still wending its way through the Senate. The USSR outwardly observed the diplomatic proprieties: it joined in the Security Council's vote to condemn the seizure; the Soviet judge on the International Tribunal in the Hague adopted the same stance. But once it became clear that the US would not in fact use force, the Soviets' tone towards America became strongly admonitory and its attitude toward Iran's outrageous action rather indulgent, along the lines of "of course we deplore it, but we can understand why it was done."

This barely concealed satisfaction comes out clearly in an authoritative Soviet pronouncement on the subject. "It is indisputable that, taken by itself, the seizure of the American Embassy cannot be justified in view of the international convention concerning diplomatic immunities and privileges. One must not, however, judge this act in isolation from the general context of American-Iranian relations, or forget about those activities of the US vis-à-vis Iran that in no sense can be reconciled with [international] law and morals."[33] The article goes on to report disquieting rumors about the Pentagon's designs to bomb Iran's oil installations or even use atomic bombs. "Whatever the facts, it is clear that we are witnessing blatant military and political pressure on Iran. . . . The US . . . is resorting to blackmail in regard to a sovereign state. Instead of showing prudence, restraint, and caution . . . [and] striving for a reasonable resolution of the problem . . . certain circles in the US are more and more inclined to resort to force." How about America's own past and present violations of international law in regard to Iran—e.g., rejecting the Iranian people's demand that the Shah be turned over to them? Here, as not infrequently, the Soviet commentator could quote some supporting voices from the US, such as an American senator who allegedly said that the "Shah presided over one of the most cruel regimes in the history of mankind," and a former

33. A. Petrov, "For Prudence and Restraint," *Pravda*, December 5, 1979.

American UN representative who had contributed to the discussion by declaring, "The US is sheltering a murderer and robber." Certainly, adds our Soviet writer, considering his crimes, the surrender of the Shah to Iranian justice would not be incompatible with American national honor and reputation. The article concludes by once more drawing attention to the inflexible principles of Soviet foreign policy: "strict respect by all states, whether big or small, of the sovereignty of other countries, non-interference in their internal affairs, observance of the established norms of international relations, diligent search for peaceful solutions of all kinds of conflicts and problems. . . ."

The temptation to gloat over yet another American discomfiture was too strong to resist. And yet the USSR had seemingly little to lose and possibly a great deal to gain by condemning unqualifiedly the embassy seizure. A public message, say, from Gromyko to Vance expressing outrage at the act and sympathy for the US in its predicament would have gone far in smoothing the passage of SALT II through the Senate, and in tempering American indignation when, shortly afterwards, the Soviets themselves put a peculiar interpretation upon the phrase "respect . . . of the sovereignty of other countries [and] non-interference in their internal affairs." Conceivably the Kremlin believed that if it tried to humor Washington, the latter might be emboldened to take a tougher line. With the presidential elections approaching and Carter being challenged from the left within his own party, Moscow evidently concluded that his administration, come what may, could not afford to weaken its commitment to detente or abandon its efforts on behalf of SALT II. The Soviet press depicted the American scene as one of confusion and disarray, giving special attention to Senator Kennedy's attacks on the US's past record in supporting the Shah and to Carter's declaration that he would eschew the hustings and stick to the White House as long as the hostage crisis continued. A government so indecisive and politically vulnerable—one that would not at first grant asylum to a fallen ally and then became apologetic about admitting him when he was mortally ill, only to embark immediately upon shamefaced attempts to get rid of him—was not likely to react vigorously to a new act of force by the USSR.

On December 25–26 a massive airlift poured thousands of Soviet troops into Kabul. Simultaneously, armored units that had been stationed on the border for some time moved into Afghanistan, along the modern highway the USSR had thoughtfully built for its southern neighbor several years before the coming of the Communist regime. On the twenty-seventh the Soviet troops struck at and occupied key points in the capital. President Amin and his family and entourage were massacred.[34] The new Soviet satrap in Afghanistan was Babrak Karmal, who succeeded to the murdered Amin's governmental and party offices. This leader belonged to the so-called Parcham faction of the Afghan Communist movement, which in the past had feuded with the dominant Khalq group of Taraki and Amin. Dismissed from his govenment position shortly after the April 1978 coup, Karmal had found shelter in Czechoslovakia and evidently arrived in Kabul along with the Soviet troops, a fact that did not prevent official USSR and Afghan sources from asserting that he had been the leader of the "patriotic forces" that overthrew Amin and then asked the fraternal Soviet people for the military help that materialized with such rapidity.

Soviet dissatisfaction with Amin dated from the first days of the Afghan Communist regime. Moscow believed him to be primarily responsible for the doctrinaire measures undertaken by the new government that had ignited the countrywide insurgency that Kabul proved incapable of suppressing. Visits by high-ranking Soviet military and security officers confirmed Moscow's judgment that Amin had to go before the regime could hope to achieve any kind of stability, not to speak of popular acceptance. This was clearly the advice given to Taraki during a visit to the USSR in September. Upon his return to Kabul, "the Great Leader," as he was called in the official Afghan press, hastened to accommodate his Soviet friends. One version has it that Taraki, with the assistance of

34. Anthony Arnold, *Afghanistan* (Stanford, 1978), p. 96.

the USSR ambassador, lured his second-in-command into an ambush in which several of the latter's escort were assassinated but Amin himself escaped.[35] Whatever the details, Amin then turned the tables on his treacherous boss.

The killing of the faithful henchman who only a few days before had been featured in the Soviet press as conferring with Brezhnev could not be very pleasing to the Kremlin. But, never sentimental in such matters, the Soviet leadership accepted for the moment the *fait accompli.* On September 18 Brezhnev and Kosygin dispatched a public message to "Comrade Hafizullah Amin" in which they congratulated him on his "election" and expressed "the conviction that in the future *too* the fraternal relations between the Soviet Union and revolutionary Afghanistan will continue to develop on the basis of [their] treaty of friendship, good neighborliness, and cooperation."[36] One wonders why the Soviets did not intervene then and there, but also why Amin did not take special care to deter the USSR from offering yet another proof of its good-neighbor policy.

Perhaps Moscow would have been willing to tolerate Amin had he shown any sign of being able to master the insurgency. But he continued policies that were alienating a wide spectrum of the population. The military situation was worsening, with entire units of the regular army going over to the rebels. It would have been a "first," and a bad precedent: a Communist regime closely allied to the Soviet Union overthrown by a popular uprising.

The exact details of the Soviet December 25–27 coup remain unclear, but apparently Amin died fighting, and units of the army loyal to him offered resistance to the invader. On January 3, the Soviet press announced "the untimely death" on December 28 of Gen. Victor Paputin, a Soviet deputy minister of the interior, who had paid an official visit to Kabul one month before and stayed for more than two weeks. He then presumably returned to help with the invasion and died in the fighting.

"Comrade Amin" of September 1979 became as of Decem-

35. *Ibid.,* p. 84.
36. Quoted in *Pravda,* September 18, 1979. My italics.

ber 29 "a bloodthirsty agent of American imperialism, oppressor and dictator . . . murderer, charlatan of history [sic!]." Such at least was the epitaph pronounced upon him in a radio broadcast by Babrak Karmal, approvingly reported by *Pravda*. "Overthrown by a wave of popular indignation, the treacherous scoundrel was tried and shot." And now Karmal in a radio broadcast stressed eloquently the theme of national unity and reconciliation: "I greet you, my long-suffering compatriots, worthy Moslems of Afghanistan, Sunnis and Shiites, ulemas [Moslem teachers] . . . soldiers and officers . . . national tradesmen and entrepreneurs, patriotic landowners, hard-working craftsmen, Afghan tribesmen . . . shepherds and nomads, government officials, workers and peasants, progressive youth, the intelligentsia, fathers and mothers, brothers and sisters . . . ; all of you who until now suffered under the yoke of this murderer . . . and his henchmen."[37]

Though presumably all that could be said on the subject had been said, the next day *Pravda* gave Soviet readers some further details. Amin, it turned out, had been a tool in the hands of "the imperialist circles of the US, Beijing's leaders and governments of some other countries. Through deceit and intrigues, he had seized the levers of power and then dismissed and physically annihilated the lawful President, Noor Taraki. Through his evil deeds, gross violations of legality and order, cruelty, and abuse of power, he undermined the ideals of the April revolution."

The tyrant, it was given out, had been overthrown entirely by domestic forces, but "under the circumstances, the [Afghan] government turned to the Soviet Union with an urgent request for immediate help to fight external aggression. The USSR decided to accede to the request and to send into Afghanistan a modest military contingent that would be used exclusively in fighting armed interference from the outside. . . . The request of the Afghan government and the positive response of the Soviet Union are also based on Article 51 of the United Nations Charter, which proclaims the inalienable right of states to self-defense . . . to thwart aggression and

37. Quoted in *Pravda*, December 30, 1979.

restore peace." Such aggression, the piece stated, was coming from the Afghan refugee camps in Pakistan; the CIA and Beijing were its main instigators. The article, "Concerning the Events in Afghanistan," carried a signature indicating an authoritative Kremlin spokesman.

Though the contingency plans and preparations for the invasion had been in the works for a long time, the decision to strike does not seem to have been reached before some time in November. There might well have been dissenting voices within the leadership. No Soviet expert on the US could have failed to warn the Politburo that this step would almost inevitably bring the defeat or postponement of SALT II, and possibly place detente as a whole in jeopardy. But by now the braking system within the Soviet foreign policy apparatus had for too long been in disuse. The argument that prevailed must have stressed that the die had been cast in April 1978. The USSR could not now afford to let a Communist regime in a neighboring, strategically important country be destroyed. And was it entirely improbable that, in his predicament, Amin might have turned for support to Beijing or even the US? The Soviets' fantasies as to what has actually happened are not infrequently a good indication of their fears of what could take place in the future.

Though undoubtedly forewarned, the Soviet leaders must still have been astpunded by the intensity of the American anguish over Afghanistan. For all their abundant experience on that count, the importance that public opinion in this country attaches to the legal and moral aspects of international affairs has not ceased to baffle the Kremlin. Why does the US become much more concerned when Russia uses force to preserve its domination over a country than when it acquires such domination in the first place? Why are Americans, themselves so prone to blame the USSR for all the troubles on the world scene, so perturbed by a few harmless fibs of Soviet propaganda, e.g., that Amin was a CIA agent? How pedantic and incongruous for Washington to consider unattractive characters like Pol Pot and Amin as defenders of national sovereignty!

No American figure epitomized these incomprehensible (to the Kremlin) attitudes more strongly than President Carter. Following the invasion, he hastened to contact Brezhnev on the "hotline" and asked him to pull the Soviet troops out. The Russian leader patiently rationalized his government's actions in terms with which we are already familiar. In his televised interview on December 31, the President in effect characterized this answer as a falsehood and declared that the events of the last few days had completely changed his view of the Soviet Union and its goals.

In 1977 Gromyko, defending the USSR from charges of interfering in African states' internal affairs, could assert that on the whole continent there was not "a single Soviet soldier with a rifle." There were now quite a few such soldiers in Afghanistan; "the modest military contingent" grew by the end of January 1980 to 80,000–90,000 troops. On January 3 the US President requested the Senate to postpone indefinitely consideration of the SALT II treaty. On the fourth Carter announced other measures directed against the Soviet Union. The most important was the curtailment of grain sales. The US would honor its obligation under the Soviet-American trade treaty to deliver a specified quota of foodstuffs, but stopped the shipment of the additional seventeen million tons contracted for by the USSR because of its poor 1979 harvest. A ban was placed on the delivery of high-technology items, including oil-drilling equipment and computers. The US also announced it would probably refrain from participating in the Moscow Olympic Games scheduled for the summer of 1980. There were other punitive measures of lesser importance.

Washington's indignation and protest were echoed in other Western countries and by most of the non-aligned states. On January 14 the United Nations, by a vote of 104 to 18, called for the immediate withdrawal of "the foreign troops from Afghanistan." A number of Moslem countries, including Algeria, Syria, and Libya abstained from voting or chose to be absent; India also was absent. Of the Soviet-bloc countries, Romania was absent. The Indian representative informed the assembly that his government had been assured by the USSR that the troops would "be withdrawn as soon as possible." "When will you stop killing people?" asked an indiscreet British visitor of Stalin in the 1930s. "When it is no longer necessary," calmly answered the Soviet leader.

What made Moscow's move particularly disturbing to many was that this was the first time that Soviet troops, as distinguished from surrogates, had been used for aggressive purposes outside Eastern Europe—another example of that Western punctiliousness which must both irritate and amuse the Kremlin.

Gradually the usual pattern of developments following a Russian act of aggression began to reassert itself: the Soviet presence in Afghanistan continued to be deplored, but increasingly accepted as a fact of international life—one which, many argued, should not be used as a pretext for reviving the Cold War. America's allies continued to denounce the invasion, but showed little inclination to follow the US example by imposing meaningful economic sanctions. Of the latter, the one potentially most damaging was largely offset by expanded Soviet grain purchases from countries like Canada and Argentina. The most spirited discussion touched on perhaps the most trivial issue: whether or not to participate in the Moscow Olympic Games. Some argued that the world fraternity of athletes must not be disrupted or penalized on account of politics, since sports and international competition were a great force for international understanding and peace. Opponents of participation tended to carry their side of the argument to rather unreasonable lengths, as in the assertion that had the democracies boycotted the 1936 Berlin Olympics, Hitler would have been so chastened that World War II might have been avoided.

It was in the Moslem world that one could have expected the Soviet invasion to have a special impact. But as we have already suggested, this impact was hardly of the kind to fill the Kremlin with trepidation. It was significant in this context that on one occasion, when a crowd of Afghan refugees and Iranian sympathizers advanced menacingly on the Soviet Embassy in Teheran, the authorities took prompt steps to head

off the demonstrators and to protect the building and its occupants. None of the Moslem countries that had diplomatic
relations with the USSR felt it advisable to sever them in the
wake of the invasion. Fear was proving more potent than a
sense of outrage and Moslem solidarity. Pakistan, to be sure,
was sheltering the Afghan refugees and freedom fighters, but
not without serious apprehensions. The long arm of Moscow
now extended quite close to the Persian Gulf; the Soviets were
in an even more favorable position than before to exploit and
exacerbate the ethnic divisions in Iran and Pakistan.

Even in the United States the initial and widespread approval of the administration's response to Afghanistan was
soon tempered by some second thoughts and criticism. Business firms trading with Russia saw their canceled orders being
filled by their German, French, etc., competitors. Farmers
complained that it was unfair to place upon them the main
burden of Washington's resolve to teach the Soviets a lesson.
Voices within what had become known as the arms control
community, as well as many others, questioned the logic of
linking Soviet misbehavior with the fate of SALT II. Besides,
it was being argued, what was the point of sanctions if they
did not work? They were obviously not going to get the Soviet
troops out of Afghanistan. The theme of a "Soviet Vietnam,"
brought up before in connection with Angola, Ethiopia, etc.,
surfaced once again: the US had no reason to become overly
concerned simply because the Russians got involved in yet
another Third World country. This involvement would place
additional strain on their already overextended military commitments and economic resources, and the enterprise itself was
in the long run self-defeating. Such adventures were bound to
make the Soviets weaker, hence in the long run more tractable.

Despite such rationalizations and consolations, few in this
country could fail to share President Carter's shock and feeling of having been deceived in his hopes concerning the USSR.
For more than a decade Washington and American public
opinion had tended to give the Kremlin the benefit of the
doubt: for all of its peccadilloes in Africa, the main goal of

Soviet foreign policy had been to reach an accommodation with the West. Now the conviction grew that detente, in Moscow's scheme of things, was secondary to—indeed one of the means towards—the achievement of its imperialist aspirations. With East-West negotiations apparently unavailing, what means could be devised to stop the momentum of Soviet expansion?

6

At the Crossroads

To all appearances, detente was dead or moribund. But in January 1980 this diagnosis would have been vigorously disputed by the Kremlin and thought at least premature by some Americans, as well as a much larger number of Western Europeans. As originally conceived and as its common-sense definition implied, detente meant a relaxation of international tension, resulting primarily from the US and the USSR having reached an agreement on strategic arms control and adopting other measures and subscribing to principles designed to assure peaceful coexistence. Now international tensions had perceptibly risen, SALT II was in abeyance, and American and Soviet notions as to what constituted peaceful coexistence were clearly miles apart.

But the Soviets seldom discard political concepts judged useful to their purposes just because such concepts have ceased to correspond to reality. The Soviet Union, Moscow insisted, still adhered punctiliously to the letter and spirit of the Nixon-Brezhnev agreements and declarations of 1972. It was then not the Soviet policies but those criticizing them who were undermining detente. As Brezhnev put it in a declaration pub-

lished by *Pravda* on January 12, "Today the opponents of peace
and detente are trying to capitalize on the events in Afghanistan . . . a shameless anti-Soviet campaign is being mounted."

Such sleights of hand were not without their effect on public opinion in the West. Whatever its original meaning, detente had come to be widely seen as the most hopeful alternative to sharply hostile relations between the two camps, a
concept to be clung to because as long as both sides acknowledged that it was still in force, confrontation and war were
unlikely. It is thus understandable why the West's initial reaction of outrage over Afghanistan soon gave way to a feeling
of relief that this strong condemnation of the Soviet actions
did not provoke Moscow to announce the demise of detente,
and that on the contrary, the Kremlin remained, in Brezhnev's
words, "strongly in favor of consolidating and multiplying
everything positive that has been created over the years on the
European continent."

As these words plainly indicated, the USSR sought to spotlight and exacerbate the differences that had developed between the US and its European allies. The US, Brezhnev also
stated, had been shown to be an "absolutely unreliable partner" in international affairs—a characterization that came close
to what quite a few Western Europeans had in recent years
felt and on occasion even said about American foreign and
economic policies. Very astutely the Soviet leader referred extensively to SALT II, currently placed in cold storage by the
Carter Administration. The whole world had fervently desired
SALT II to come into force, for it would have opened new and
tantalizing vistas of international cooperation and security; "it
would have cleared the path for really important progress in
disarmament." But "hardly had the treaty been signed when
. . . not without the connivance of government circles . . .
some people in the US sought to discredit it." Brezhnev could
not have been unaware of the argument heard in Washington
and other NATO capitals that the current crisis was partly of
America's doing: had the Senate not been so dilatory about
ratifying SALT II, the Russians might not have been so precip-
itous about Afghanistan. But with the treaty quite likely to be

rejected or, if ratified, burdened with crippling amendments, the Soviets allegedly felt they had little to lose by resorting to military action.

While trying to deepen the fissures in the Western alliance, the Kremlin did not wish to revert to a Cold War posture vis-à-vis the US. But rather than mollifying the Americans, it hoped to teach them a lesson: pressure cannot deflect the USSR from its course once it has settled on a definite policy; your own allies will not follow your lead, and by persisting in your anti-Soviet policies you place the alliance itself in jeopardy. There was also an element of personal pique at Washington's unpredictability. It was not only for the benefit of Western Europeans but out of a profound sense of irritation that Brezhnev described the American leadership as prone to change its policies on the basis of "some whim, caprice, or emotional outburst." The US was proving incapable of handling the Ayatollah and protecting its own citizens, yet it thought it could frighten and punish the Soviet Union!

Of the sanctions invoked by Carter in reprisal for the Kabul coup, the grain embargo was the one least anticipated and most acutely felt by Moscow. It must have appeared almost inconceivable that an administration that was already reeling under the impact of the hostage crisis and facing presidential elections only a few months away could summon enough courage to take a step that was bound to antagonize an important segment of the American electorate. With their own economy experiencing serious problems, and anticipating their second bad annual harvest in a row, the Soviets were especially sensitive to any foreign pressure that sought to magnify and capitalize on their domestic difficulties. Gross agricultural production fell by 5.8 percent in 1979; it would decline 4.4 percent further in 1980.[1] In his January 12 message Brezhnev tried to reassure the home audience by pledging that even though the US had reneged on its contracts for grain deliveries, the "plans for providing the Soviet people with bread and bakery products will not be reduced by a single kilogram." But this bypassed

1. CIA National Foreign Assessment Center, *Handbook of Economic Statistics, 1981* (Washington, D.C., 1982), p. 57.

the real difficulty: because of the decline in the stocks of animal feed, the Soviet consumer would eventually have to cope with serious meat shortages.

When it came to the Afghan venture, the Kremlin had expected—up to now correctly—that its impact on the domestic scene would be minimal. The Soviets had had ample experience in dealing with guerrilla warfare, most notably in their own Central Asia in the 1920s, when the Basmachi (the name given to the indigenous rebels against Communist rule) held out against the Red Army, in some areas for years. In Afghanistan, Moscow did not duplicate the American tactics in Vietnam. Instead of pouring in hundreds of thousands of troops in the hope of decisively and quickly defeating the insurrection, the Soviet army's principal role in the country had been to preserve and bolster the Communist regime, as well as to keep it under tight control, so as to prevent it from spawning new Amins. The Russians guarded the main cities and strategic routes, occasionally applying their superior military technology to eradicate a nest of rebels, trying at the same time to build up the regular Afghan army into a reliable tool of the regime. With the Soviet garrison estimated at about 100,000, the troops being frequently rotated, and casualties on the order of 10,000 per year, the military aspect of the Afghan enterprise had thus not been inordinately costly to Moscow, or politically damaging on the domestic front.

Where the USSR had clearly miscalculated was in not foreseeing the intensity and tenacity of popular resistance to the satellite regime. Babrak Karmal's Soviet masters had expected that the new leader's discarding of his predecessors' drastic social reforms, his wooing of the Moslem religious community and posing as a moderate and nationalist (as evidenced in his speech quoted above) would go far toward pacifying the turbulent country and gaining a measure of popular acceptance for the regime. Yet despite such posturings in 1982, more than two years after the invasion, all the indications were that the Kabul government remained as detested as ever by the majority of the population. Partisan struggle in the countryside had not slackened, and even in the capital and main cities the pro-regime and Soviet personnel had to take special precautions if they ventured out of doors after dark. More than a million Afghans had sought refuge in Pakistan.

Afghanistan is far from having become "Russia's Vietnam," and other crises have preempted the world's attention; but the unfortunate country has remained a reminder of one of the Soviet Union's less successful imperialist ventures, and a potential focal point for future troubles and conflicts.

National pride—that not inconsiderable element in the Soviet outlook on international affairs—had to be somewhat affected by the partial boycott of the Moscow Olympic Games. Like a football hero eager to display his prowess and sophistication at a school dance, the Soviet regime was anxious to prove to its people, and especially to the young, that their country was great and in keeping with the times, not only because of its military and industrial power, but also on account of its government's solicitude and skill in providing for the lighter, but so important, aspects of modern life. The Olympics were to demonstrate that Russia had also come of age insofar as the amenities of civilized existence and leisure were concerned and that Moscow need no longer fear invidious comparisons with Western capitals on the part of the hundreds of thousands of foreign tourists who visited it for the games. The fact that for the first time in history the spectacle was scheduled to take place in Russia was a flattering recognition not only of Soviet preeminence in sports—that had been established for some time—but also that socialism, no less than some effete capitalist systems, was able to provide an environment of comfort, elegance, and gaiety appropriate for such extravaganzas.

And so in anticipation of the summer 1980 events, new hotels and athletic complexes sprouted up in and around Moscow. The ancient capital was being spruced up and stocked with provisions to accommodate the crowds of visitors. The police and the KGB were busy compiling local lists of already and potentially vocal dissidents who would be invited to reside elsewhere for the duration.

A cancellation of the games would have been to these pas-

sionately sport-loving people the most tangible and direct
demonstration of how the world viewed their rulers' actions
in Afghanistan. A widespread boycott had to be seen by the
Soviet establishment as a potential propaganda setback quite
out of proportion to the importance of the event itself. There
must have been sighs of relief in the Kremlin when most
Western Olympic committees and athletes refused to heed their
governments' remonstrations and chose to participate. The
games were held with more than eighty nations taking part.
The occasion was a bit soured by the absence of such athletic
powers as the US and West Germany, and foreign tourist at-
tendance was smaller than it would otherwise have been. But
the Muscovites' pride had been salvaged; their city had hosted
a prestigious international spectacle.

The failure of the sports boycott was symptomatic of the
growing inability of the West to implement any concerted plan
of action vis-à-vis the USSR. The governments of the Continent's principal powers were emphatic in their *declarations*:
the Soviets must pull their troops out of Afghanistan. At the
same time, Germany's Chancellor Helmut Schmidt and
France's President Giscard d'Estaing continued stressing the
importance of not abandoning the West-East dialogue and arguing that economic sanctions such as those adopted by the
US were essentially counterproductive and likely to lead to
the hardening of the Soviet position, etc. In their speeches, the
Western European statesmen acknowledged that detente had
to be global in nature to retain any meaning, but their actions
bespoke a determination not to antagonize Moscow as long as
Europe itself or its interests (such as its supplies of energy)
were not being directly threatened by Soviet actions. The British government, now under the Conservatives, found itself
closer to the American position; its foreign secretary declared
that "the West itself needs to find ways to make Russia understand it cannot break the rules of international behavior with
impunity, either now or in the future."[2] But Britain's weight,
whether politically or economically, was now considerably

2. *The Times*, January 25, 1980.

below that of West Germany and France in the councils of Europe.

America's allies could see through Moscow's game; it was trying to play them off against the US, thus weakening the bargaining position of both. Western Europe, Chancellor Schmidt repeatedly declared, could not remain an "oasis of detente" while detente was not being observed in the rest of the world. Such reservations notwithstanding, the Chancellor visited Moscow in late June 1980. There he approved the preliminary plans for a vast joint Soviet–West German commercial undertaking. The Federal Republic would help construct and finance a pipeline to bring Siberian natural gas to West Germany. When completed the pipeline would account for a sizeable proportion of the latter's energy needs. Apart from thus acquiring a potential means of pressure on Bonn, the Soviets would garner very handsome earnings in hard currency from the deal. Chancellor Schmidt's plea to his hosts that they commit themselves to an *eventual* complete military withdrawal from Afghanistan fell upon deaf ears, and the relevant passages of his speech were omitted in the Soviet press reports.

Viewing the overall world situation, one might have said that the era of containment had returned, but with a difference. It was the United States that for the moment was being contained; its own allies were rebuffing, in deeds if not words, its policies vis-à-vis the USSR. One had to marvel at yet another tour de force performed by Moscow's diplomacy: in the wake of Afghanistan, it was not Soviet actions but the American reactions to them that appeared to many to endanger peace and detente.

At the same time, by trying to exploit to the hilt the discomfitures of American foreign policy, the Kremlin seemed to abandon any restraint and to forget that it had to deal with the current Democratic Administration, at least until January 1981. Its at least verbal recklessness was well illustrated by the Soviet reactions to the abortive US attempt to rescue the American hostages in Iran. The news that an American combat team that had landed on April 25 in a desert area of Iran had encountered technical difficulties that led to the scrapping of the mission, with eight servicemen accidentally killed, was reported in the Soviet press the next day. Usually some time elapses before the Soviet press receives instructions on how to interpret an event of this kind. But now the reaction was instantaneous and scathing. Its two main themes stressed President Carter's personal responsibility for the "dangerous military-political adventure [that] has ended in a disgraceful fiasco" and the danger to which Washington's adventuristic policies exposed its allies and indeed the peace of the world. Western Europe was called upon to help restrain the Carter Administration's "irresponsible game." Again Moscow sought to capitalize on the Europeans' apprehensions about the unsteady course of US foreign policy: "Thus the 'complaisance' of the West European allies has led to results quite opposite to those which were expected. The policy of 'appeasement' has made the American President even more impulsive, adventurous, and unpredictable. This policy threatens to draw the West European countries, against their will and contrary to their interests, into a dangerous conflict in the Middle East."[3]

The extent to which the unfortunate raid was exploited by the Soviets for personal attacks upon Carter must in retrospect be judged as somewhat puzzling. From this side of the ocean it appeared that the President had been very patient and accommodating in his dealings with the Russians. The Kremlin's experts on American affairs must have advised of the possibility, if not probability, that any alternative administration would take a much harsher line vis-à-vis Moscow. But for the Soviet policy-makers, the short-term advantages in exploiting the West European leaders' well-known personal reservations about Carter evidently took precedence over considerations of how they might fare with a new president in the White House.

As on many previous occasions, the Kremlin felt less con-

3. "Commentator's Opinion," *Pravda*, April 26, 1980.

cerned about the political complexion of any leadership in Washington—current or prospective—than about its own ability to gauge and predict its policies and reactions. And by now Brezhnev and his colleagues had decided that Carter was quite unreliable. *Pravda* expatiated on the theme: "It is not without reason that the term 'unpredictable' has been appearing in the newspapers more and more when US policy is discussed. Some people are trying to attribute this unpredictability to the upcoming presidential elections, which, they say, are forcing 'poor Carter' to resort to drastic measures. And even if that were the case, isn't Carter willing to pay too high a price to stay in the White House for another four years? And isn't 'unpredictability' in the actions of the head of a state such as the US too dangerous in the age of nuclear missiles? These questions are being asked with increasing frequency in the West."[4]

The resignation of Secretary Vance on April 28 led to further personal attacks on the President. Vance was praised not only because he had opposed the hostage rescue attempt, but also as one American statesman "who favored, although not always with the proper persistence, the maintenance of detente." His departure was described as "a serious blow to the adventuristic policy of Carter and Brzezinski," for "it revealed that this policy is causing growing apprehension even among certain segments of US ruling circles."[5]

Not since 1949, when the Moscow stage presented *The Mad Haberdasher* (a subtle reference to Harry Truman's one-time occupation),[6] had the Soviets engaged in such a systematic and vituperative campaign against an American president. Had Carter been reelected, the Russian media undoubtedly would have changed their tune. But at the moment, he was seen as politically weak and vulnerable, hence the irresistible urge to

4. Yuli Yakhontov, *Pravda*, April 27, 1980.

5. Yuri Zhukov, *Pravda*, April 30, 1980.

6. And a testimony to the Soviets' snobbery. Why, and especially from the proletarian point of view, should having been a haberdasher disqualify a man for high office?

indulge in vilification of the man whose outspokenness on the human rights situation in the USSR had touched a raw nerve of the Soviet establishment.

Relations with the US were then placed on "hold" while the Kremlin intensified its propaganda campaign in Western Europe. Its primary aim was to create there a climate of public opinion that would make it virtually impossible for the respective governments to follow the American lead in any policies opposed by Moscow. On April 28–29 a meeting of European Communist parties was convened in Paris, designed to spotlight the main themes of the Soviet campaign. The head of the Soviet delegation, alternate Politburo member and secretary of the Central Committee, Boris Ponomarev, delivered his charge to the conference: "European Communists should place themselves at the head of a broader popular movement that must check the dangerous drift away from detente and toward war." Specifically, the Europeans must try "to frustrate NATO's dangerous plans to deploy new American nuclear missiles on our continent, . . . to prevent a new round of the arms race . . . and the sabotage of the entire course of peaceful collaboration between the socialist and capitalist states of Europe, . . . [and] not to allow the structure of exchanges in the fields of science, education, and sports that was established in the 1970s to be upset." In brief, Western Europe must not be allowed to be an accomplice of US imperialism, "a role which is at variance with the national interests of the West Europeans and threatens to involve them in adventures far from home." [7]

Though this post-Stalin international Communist gathering could not be expected to go completely smoothly from Moscow's point of view—and in fact its representatives were exposed to some sharp criticisms concerning Soviet policies in Afghanistan—the meeting as a whole endorsed the main points of the Soviet appeal. The joint declaration evoked memories of the Popular Front of the 1930s as it called for a broad alliance of "Communists, Socialists, Social Democrats, Christians, and people of other persuasions"; the alliance was

7. *Pravda*, April 29, 1980.

to strive for the cancellation of NATO's decision to install intermediate-range nuclear weapons in Europe and for the "earliest possible ratification of SALT II."[8]

The new pattern of Soviet diplomatic strategy could thus be seen quite plainly: the Russians were working not so much to break up the Western alliance, but for the European partners in the alliance to become a pressure group that would prevent the US from developing any independent initiatives vis-à-vis the USSR. According to the Soviet scenario, whoever occupied the White House in January 1981 and whatever his intended policies, he would find his hands largely tied, with Bonn, Paris, and other NATO capitals pressuring Washington to retreat to the pre-Afghanistan attitude, ratify SALT II, and modify its position on intermediate-range nuclear missiles. Under such circumstances NATO would be transformed from an organization designed to block Soviet expansion into something more like a debating society, with its internal disputes and squabbles inhibiting its members from developing, whether jointly or singly, any effective countermeasures to the Kremlin's moves.

And to strengthen the chances of this scenario becoming a reality, the USSR would, at least for the time being, observe restraint and moderation in its behavior.

For many in the West, a test of such restraint came in May 1980, following the death of Marshal Tito. As the famed leader of Yugoslavia, who had been seriously ill for some time, approached his end, many throughout the world awaited fearfully the Soviet's reaction to his disappearance from the scene. Both on account of the legend already surrounding him and through the force of his personality, Tito had succeeded even in extreme old age in keeping his politically troubled country united and independent. The Kremlin, it was known, had never quite acquiesced in Yugoslavia's 1948 defection from the Soviet camp, the first harbinger of that spirit of revolt against Moscow's absolute domination of the Communist world that became so pronounced after Stalin's death. Tito and the Soviets were officially reconciled in 1955, but the latter never gave

8. *Izvestia*, April 30, 1980.

up trying, whether by internal subversion or diplomacy, to bring Yugoslavia back into the fold. Both intrigues and cajoleries were stoutly resisted and fended off by the Marshal, who was as skillful in manoeuvering between East and West in foreign policies as in alternating between reform and repression in domestic ones. But already during Tito's last years, Yugoslavia's internal problems became exacerbated, placing in jeopardy the country's unity and stability: ethnic conflicts and political dissent were compounded by growing unemployment, and the country's economy, tied closely to that of the West, also suffered from recession.

And now, with his personal magic gone, it was believed problematic whether Tito's successors would be able to preserve his legacy. The regime he bequeathed to them could please neither a liberal nor a Communist true believer. It aroused resentment among the Serbs, since they no longer had the dominant position in the state, at the same time that many Croats and Albanians felt that their national aspirations were still being slighted. Though most Yugoslavs of all nationalities still would not have wished to exchange their country's status for Czechoslovakia's or Bulgaria's, there were pro-Soviet elements among the malcontents. And so it was not out of the question that with Tito gone, some within the party and military elite might opt for a closer link with the Soviet Union as one way of preserving Yugoslavia's unity and its Communist regime.

We can now see that the fears of Yugoslavia reverting to the status of a Soviet satellite were greatly exaggerated. And it was unreasonable even to suggest, as Western observers frequently did, that the Soviets would resort to an invasion. Had conditions approaching civil war developed within Yugoslavia, and had one of the factions called for "fraternal help," the USSR might have been sorely tempted to oblige. But short of such a drastic eventuality, nothing indicated that the Kremlin would as much as consider a fresh military adventure.

The Soviets went out of their way to quash rumors of any possible interference in Yugoslavia's affairs. The deceased leader was eulogized in both official Soviet government statements and the press as a towering figure in the international Communist movement. Newspaper accounts of his career glossed over the period 1948–53, when no term in Moscow's formidable arsenal of abuse was spared to describe the man who dared to defy Stalin, and when "Titoism" took its place alongside "Trotskyism" as a synonym for the extreme in treachery and villainy. Omitted from the obituary notice were any references to other periods of tension between Yugoslavia and the USSR; even after the formal reconciliation had taken place, each time fresh troubles struck their empire—in Poland and Hungary in 1956 and Czechoslovakia in 1968—the Soviets had second thoughts about having "forgiven" Tito his heresy, and Moscow-Belgrade relations experienced a new period of stress.

But now the highest Party and government bodies of the USSR hastened to assure Tito's heirs that "the Soviet people share the grief of the Yugoslav working people in connection with the death of Josip Broz-Tito—the outstanding state and party leader of Yugoslavia, a prominent figure in the international Communist and workers' movement, and a tireless fighter for peace. . . . In these days of mourning, the CPSU Central Committee, the Presidium of the USSR Supreme Soviet, and the Council of Ministers reaffirm the Soviet Union's invariable striving to strengthen friendly relations with the Socialist Federal Republic of Yugoslavia . . . on the basis of steadfast observance of . . . noninterference in each other's internal affairs and equality and strict respect for sovereign rights, in a spirit of mutual understanding and trust." [9]

It was a message startlingly different in tone from the one the Soviet authorities sent to China upon the death of Mao. But it may well have occurred to Brezhnev, as he stood solemnly by Tito's bier, that the man to whom he was paying homage had been the first to enter the path since followed by the Chinese Communists, and that the challenge he had epitomized still confronted the Soviet state and system with its gravest dilemma.

Within a few months of Tito's passing, the notion of a So-

9. *Pravda*, May 6, 1980.

viet invasion of Yugoslavia must have appeared to the Soviet
leaders as a bad joke. For now it was no longer a question
whether they might seek a new conquest, but whether the
USSR would *have* to use military force to preserve an old one.

The Polish crisis of the summer of 1980 took Moscow by
surprise. There had been, as we have seen, ample portents of
trouble. But Brezhnev and his colleagues had come to rely
upon Edward Gierek, the chief of the Polish United Workers'
(Communist) Party. Set in their ways, the elderly men of the
Kremlin stuck to the people they had come to know and trust,
and disliked seeing new faces in positions of responsibility,
whether at home or abroad. Hence they tended to disregard
warnings coming from several sources, including the Soviet
ambassador in Warsaw, Stanislaw Pilotovich, that Gierek's
economic policies, as well as his excessive toleration of political dissent, courted disaster. The ambassador was transferred,
and those Polish Communists who criticized their boss were
fired or demoted, while Gierek and his clique continued to
enjoy and abuse their power. On July 27, the Polish leader
joined Brezhnev and other East European Communist notables
in a working vacation in the Crimea. In Moscow the State Pub-
lishing House was preparing to issue the Russian edition of
Gierek's collected speeches and articles, a special mark of distinction for a Communist, especially a foreign one.

There had already been sporadic strikes in Poland, triggered
by the July 1 announcement of rises in the prices of meat and
meat products. By early August, labor troubles became en-
demic, and on the fourteenth a sit-down strike paralyzed the
giant Lenin shipyard in Gdansk, one of the main focal points
of the 1970 workers' riots. Lech Walesa, a veteran of that previous encounter between workers and regime, now made his
appearance among the strikers and was immediately acclaimed their leader. The strikes had turned into a revolution.

To be sure, it was an unprecedentedly peaceful and orderly
revolution, which in its entire course, until the proclamation
of martial law in Poland on December 13, 1981, claimed fewer
victims than a minor urban riot in the US. But within the next
few weeks, practically the entire Polish working class rose

against the existing order, demanding, through strikes and mass meetings, genuinely free trade unions in the place of the old party-controlled ones, the right to strike, higher wages, and other economic concessions by the government. But over and above these professional demands, the striking workers called for a thorough liberalization of the regime, through relaxation of censorship, release of political prisoners, and recognition of "the individual's freedom to express his convictions in public and professional life." [10]

The Polish party high command was thrown into complete confusion by this avalanche-like movement of the workers, whose struggle soon was joined by peasants, students, and the professional classes. Gierek, his rest on the Black Sea rudely interrupted, tried upon his return to stem the tide, using at first the tactics that had proved successful in 1970: immediate economic concessions to the workers, combined with vague promises of future political and organizational reforms. These were accompanied by stern warnings: let no one doubt the government's determination and ability to crush anti-socialist forces that might try to exploit the crisis. A bowdlerized version of his television address was reported in the Soviet media, providing the Russian public with their first inkling that something out of the ordinary was happening in Poland: "Referring to the situation in some enterprises of the Baltic seaboard, E. Gierek noted that some irresponsible, anarchist, and antisocialist elements have tried to exploit the [isolated] work stoppages for hostile political purposes. . . . 'We shall not tolerate any action directed at the political and social order in Poland . . . in that respect nobody should expect concessions, compromises or even hesitations.'" [11] As seen above, *Pravda* used the euphemism "work stoppage" for strike, the latter being by definition something that takes place only in a capitalist country.

But, unlike the situation in 1970 when monetary concessions, soothing words, and warnings succeeded in the end in pacifying the workers, this time neither cajolery nor threats

10. Neal Ascherson, *The Polish August* (New York, 1982), p. 291.
11. *Pravda*, August 20, 1980.

could arrest the growing momentum of protest and strikes. There were voices within the party calling for the use of force, but even those in the ruling oligarchy who in December 1981 did in fact resort to repression refused at this point to sanction repressive measures against the strikers. It was perceived that in the nation's current mood, such an attempt might be suicidal for the regime and entail consequences too frightful to contemplate. The government decided to negotiate and try to gain time.

On August 31 government representatives signed an agreement with the Gdansk strikers that on the face of it spelled victory for the workers. "The Government commission declares that the Government will guarantee and ensure complete respect for the independence and self-governance of the new trade unions, both as regards their organizational structure and their performance at all levels of activity."[12] Other concessions also echoed the workers' demands: generous pay raises, a shortened work week, and additional social benefits. For the moment few gave much thought to the effect of such concessions on the already faltering Polish economy. The nation's attention was riveted on the government's concessions of freedom of expression, its pledge to release political prisoners, and its promise to "ensure that the radio will transmit the Sunday [Catholic] mass under a detailed accord with the episcopate." Equally unheard-of for a Communist country was the promise that henceforth "radio, television, the press and publications should be used to transmit a variety of ideas, views and opinions." The completeness of the regime's capitulation was only tempered by the workers' reciprocal pledge that their unions "do not intend to play the role of a political party" and their agreement—under the circumstances this sounded almost ironic—that the Communist party would continue "to play the leading role in the state." The victorious workers also declared that they did not desire to tamper with "Poland's existing system of alliances," an acknowledgment that at the moment everybody agreed that Poland must remain

12. Quoted in Ascherson, *op. cit.*, p. 289.

within the Soviet camp and the Warsaw Treaty Organization.[13]

This acknowledgment could be of scant consolation to the
Kremlin. Within a few weeks the entire fabric of a Communist
regime had come unraveled. The Gdansk compact served as a
model for similar agreements with workers in other major industrial enterprises. The Baltic seaboard strike committee became the nucleus of the free labor union, Solidarity. The latter
grew prodigiously, soon embracing upwards of nine million
members—a great majority of the country's industrial force—
while the number of those in the old party-controlled unions
dwindled into insignificance. Along with those of the workers, the long-suppressed grievances and aspirations of other
groups and professions came into the open; farmers, teachers,
students, etc., voiced their demands and sought to join Solidarity. By the end of September, the government was capable
only of fighting a rear-guard action: negotiating and pleading,
no longer daring to threaten.

There was but one effective barrier preventing this elemen-
tal force of an aroused nation from sweeping aside the panic-
stricken regime: the shadow of the Soviet Union. Neither Solidarity, nor, it is fair to say, the majority of the defenders of
the old order wished Poland to repeat the tragic experience of
Hungary in 1956 or Czechoslovakia in 1968. Both Solidarity
in its demands and the regime in its concessions had to be
mindful of one unspoken consideration: how far would the
Russians allow them to go? The Catholic Church exerted its
powerful influence and prestige to plead for moderation and
restraint on the part of both the workers and authorities. The
aged and already gravely ill Primate Stefan Wyszynski embraced, after some initial hesitations, the cause of Solidarity,
but kept warning its leaders to be sensible and realistic in their
demands. At first almost everybody kept in mind the ineluctable fact of Poland's geographic situation and the imperative
necessity of saving the nation, so cruelly tried throughout its
modern history, from another bloodbath and foreign occupation.

13. Quoted in *ibid.*, p. 288.

Yet for all such cautions and apprehensions, the great majority of the people seemed resolved, even in the face of ominous rumblings from Moscow, not to surrender the rights and freedoms that had already been won. As the "Polish August" passed into the fall and early winter, the Soviet reactions to it were being watched by the West with increasing wonder, and by the other Communist regimes in Eastern Europe with mounting anxiety. What had already taken place in Poland was clearly more damaging to the cause of Communism both within the country itself and on the world stage than the events in Czechoslovakia that had prompted the Soviet armed intervention of 1968. And yet the Kremlin stayed its hand and appeared undecided as to what course it should follow. Would the USSR really tolerate a free Poland, with the Communist regime serving just as window-dressing for a pluralistic society? And how long would even this superficial veneer of Communist rule be tolerated if the Polish people became firmly convinced that the Soviets would not intervene under any circumstances? And how about the domino effect—wouldn't the Czechs, Hungarians, etc., be tempted to emulate the Poles and turn against their own Communist rulers?

However, there were cogent reasons for what appeared to the outside world to be the Kremlin's indecision, but in fact was a deliberate waiting game. A military intervention *at this time*, it was believed, and probably correctly, would involve prohibitive risks and costs. Unlike Czechoslovakia in 1968, the current mood of the Polish people was such that invading troops were likely to meet with armed resistance, with at least part of the Polish army joining in fighting the Russians. The Soviet army of course would prevail. But again, unlike 1968, the subsequent military occupation could not be merely symbolic in nature: sizeable Soviet garrisons would have to be stationed in practically every city of even medium size. Thus repression of this nation of thirty-five million would have to be of a character and on a scale not practiced within the Soviet empire (or at least its European part) since Stalin's time—with all that that implied in terms of the world's reaction.

The economic costs of an invasion and occupation could be

at least as troublesome as the military and political ones. The strikes, inordinate pay raises, and the shortened work week had compounded the catastrophic effects of a decade of economic mismanagement and of living on borrowed money. As it was, the USSR and other states of the Soviet bloc would have to assist Poland by providing it with hard currency, foodstuffs, and raw materials, to prevent an utter collapse of its industry and remedy the shortages of the most vital commodities. Occupation would saddle the USSR, whose own economy was experiencing difficulties, with direct responsibility for running a bankrupt country, where workers and farmers were not likely to break their backs to make the invader's task easier. Poland's indebtedness to the West now stood at over $25 billion. With Soviet troops in Warsaw it would be out of the question for the Western governments and bankers to agree to reschedule the country's payments on the debt, not to mention lending more money. In fact an invasion would in all likelihood spell the end of Western credits to all of the Soviet Union's East European protégés. And by now the economy of the entire Communist bloc was heavily dependent on the flow of credits from the capitalist world. The six Communist countries had increased their indebtedness to the West tenfold since the inception of detente: over $60 billion in 1980 versus $6 billion in 1971. Comecon would thus, should an invasion take place, face a desperate shortage of hard currency, which in turn would cripple the already faltering economies of its members and threaten the political stability of the local Communist regimes. Invasion, then, had to be viewed by the Kremlin as a dangerous gamble to be resorted to only after all other possibilities and options were exhausted, or in case the Polish Communist regime was in imminent danger of being overthrown.

The best possible eventuality, from Moscow's point of view, was for the wave of unrest and unsettling reforms in Poland to subside, allowing the Communists to recover from the shock and begin to close ranks. Once the regime gained some breathing space it could proceed, gradually, to undo the most scandalous reforms—first of all to subordinate the labor unions

again. In the meantime the USSR proposed to keep up relentless psychological and propaganda pressure, intended to bring
to their senses the hotheads of Solidarity, and by the same
token to help restore the courage and morale of the old guard
of the Polish Communist party. In the months to come, there
would be Soviet troop concentrations on the Polish border, as
well as troop movements within Poland (under the Warsaw
Pact the Soviet Union maintained two divisions as well as liaison forces within the country, but they were kept in canton-
ments, out of public sight). Stories that an invasion might be
imminent would be leaked to the foreign press, and widely
circulated in Poland. The Soviet media themselves, while acknowledging that some reforms were in fact needed in Poland, at the same time threw out dark hints about antisocialist
elements and imperialist agents having infiltrated Solidarity
and dwelt on the Polish workers' alleged impatience with those
who would undo the achievements of People's Poland.

The Kremlin also counted on the deepening economic distress cooling off the initial elation of the masses and inspiring
some second thoughts about the recently won freedoms. Outwardly the Soviet leadership preserved its calm and exuded
confidence that the Polish workers themselves would know
how to defend "socialist Poland." But behind this veneer of
official optimism one could detect an occasional trace of deep
anxiety and of impatience for some sign that Communist notions of law and order were making a comeback in Poland.

But as of the fall of 1980 there were as yet no such signs.
The Polish party was going through the tortuous process of
trying to shed its old skin, i.e., to discharge its highest officers, who would then be blamed for the mistakes and abuses
of the past. But to the people the new top layer of officialdom
appeared not very different or much more trustworthy than
the old one. Having tried, without great success, to mollify
public opinion by firing a number of Politburo members, ministers, and regional secretaries, the party finally found itself
constrained to sacrifice its highest leader. On September 6 Edward Gierek was, "in view of his serious illness, released from
his duties as First Secretary of the Central Committee and Pol-

itburo member." The Soviet press reprinted this curt announcement without adding a word of praise or any comment at all about the dismissed leader whom Brezhnev as recently as August 29 had described as his "friend and comrade-in-arms" and whom Western statesmen credited with great influence in Kremlin circles.[14] His Polish colleagues were equally unsentimental: the fiction of voluntary resignation was soon given up, and subsequently criminal proceedings were initiated against Gierek, who was accused of corrupt practices and other abuses of his office.

His successor was a veteran Communist bureaucrat, Stanislaw Kania. In his congratulatory telegram to the new captain of what was still a sinking ship, Brezhnev betrayed the Soviets' anxieties and conveyed a warning: "Soviet Communists know you as a relentless fighter for . . . the strengthening of the leading role of the Polish United Workers' Party, as one who firmly supports proletarian internationalism and the unbreakable friendship between the Polish People's Republic and the Soviet Union. . . . May the unity [of our two Communist parties] grow ever stronger . . . and continue as a crucial bond in the cohesion of the socialist alliance and of the international Communist and workers' movement."[15]

Little known before, whether at home or abroad, Kania, whose entire background had been that of a conformist and docile follower of Gierek, now struck the pose of a reformer. The game of musical chairs going on within the high command of the regime had been dubbed "renewal." Simultaneously the Communist party bared part of the depressing picture of corruption, bureaucratic abuses, and inefficiency that had brought Poland to its present straits, blaming it all, needless to say, not on the nature of the regime, but on "them"— Gierek and the other fallen potentates.

There were, to be sure, some genuine reformers within the ranks of the Polish Communists, people who believed in the possibility of reconciling Marxism-Leninism and the Soviet connection with a degree of social pluralism and political

14. *Pravda*, September 7, 1980.
15. *Ibid.*

freedom. Among the rank and file of those who remained in the party (and since August there had been a mass exodus from its ranks) there was considerable support for a genuine "renewal" of Polish socialism. But the upper echelons were still staffed, for the most part, by hardened authoritarians and conservatives, who, for all their current liberal posturings, were determined to fight tooth and nail to keep their jobs and official privileges.

Uniquely among the countries of the Soviet bloc, Poland has preserved one major institution with whose independence the state, at least since 1956, has not dared to tamper. Bitterly as they resented the influence of the Catholic Church—whose prestige had been enhanced still further by the recent election of a Polish pope—even diehard Communists had to concede that they could not fight it openly. In its present plight the regime felt constrained not merely to tolerate the Church, but to court strenuously the Catholic hierarchy. One could not imagine Brezhnev making an audience with the Patriarch of the Russian Orthodox Church his first item of business after his elevation to the position of General Secretary. But now the new Polish party and government leaders, once in office, hastened to call upon the head of the Polish Church, Cardinal Wyszynski, and after his death, Archbishop Joseph Glemp.

While seeking help and advice from this spiritual authority, the Polish notables also had to receive a blessing from the supreme political one. Early in October, Kania and the current head of the Polish government (the office of prime minister changed hands three times in the course of 1980) traveled to the Communist Mecca. Having inspected them, Brezhnev professed himself satisfied with their ability to solve "Poland's acute political and economic problems." A steady stream of Soviet generals, political figures, and economic experts was also descending on Warsaw, casting doubt on the sincerity of Brezhnev's declarations that he was confident that the Polish comrades were quite capable of handling the crisis themselves.

A parallel pilgrimage was simultaneously converging on Moscow: the leaders of other East European Communist regimes hastened there, undoubtedly to plead with their hosts to do *something* about Poland, for otherwise the contagion was bound to spread. Indeed no Communist regime, no matter how "liberal" or even anti-Soviet, could view with equanimity a phenomenon like Solidarity. The notion of genuinely free trade unions, completely free from party control, struck at the very heart of the Communist system, and that is why Romania's Ceausescu, who had stoutly opposed the Soviet invasion of Czechoslovakia in 1968, was now also conferring in Moscow. And delighted though they are whenever the Russians experience troubles with their empire, the Chinese still made disapproving noises about some features of the Polish revolution.

For the Kremlin the crisis could not have come at a more inopportune time. Over the past several years, it had been trying, not entirely unsuccessfully, to impress a double motif upon Western Europe: (1) the Soviet Union was enormously powerful, hence it was unwise for the Europeans to rely on NATO rather than on Soviet friendship for their security; and (2) Russians were essentially peaceful, hence the Europeans should give no credence to Washington's insinuations that unless NATO was strengthened the USSR might have some military tricks up its sleeve. Events in Poland cast doubt on the first proposition; despite its vast panoply of weapons, there were obvious and inherent weaknesses plaguing the Soviet bloc. At the same time the not-so-veiled threats of invasion emanating from Moscow could hardly confirm its peace-loving image. The Soviets' response to this conundrum was in line with their previous exercises in what might be called the dialectic of aggression: God forbid that the Soviet Union should interfere, militarily or otherwise, in Poland's internal affairs. But of course it could not remain indifferent if others, i.e., the "imperialists," the "anti-socialist forces" (for geographical reasons Beijing unfortunately could not be included), did interfere and thus threatened the sovereignty of the Polish People's Republic. If the Polish workers called for help from the fraternal countries, the USSR might have to comply. But this line of argument had been used too often in the past to be taken seriously even by foreign Communists. Among the latter, the Italian and Spanish parties served unambiguous notice that an invasion would mean the final breach between themselves and Moscow.

The Polish imbroglio coincided with the last phases of the presidential campaign in the US. This fact urged additional restraint and prudence upon the Kremlin. The USSR proposed to seek to revivify detente and reopen a dialogue with the US, and that would be impossible if whoever won the election was being inaugurated while Soviet tanks were rolling through Poland.

Moscow's animosity toward Carter did not abate even when the candidate of the right wing of the Republican Party emerged as his chief rival for the presidency. In fact the Soviets could not conceal their disappointment at the President's renomination by the Democratic Convention. "Though the partisans of Jimmy Carter succeeded in stemming the movement toward a so-called open convention, . . . the results of the balloting show clearly the extent of dissatisfaction, even among the Democrats, with the policy of the present occupant of the White House." And even after it became clear that, in view of Senator Kennedy's withdrawal, Carter would be renominated, *Pravda* reported that "Madison Square Garden continued to witness mass demonstrations against the domestic and foreign policies of the White House." [16]

Nor was Carter spared when he was actually renominated: "His acceptance speech was permeated by a militant spirit." The President dwelled on the need for putting an end to the nuclear arms race but, noted *Pravda*, "Such professions have been given lie by the activities of his own administration—its heightening of international tensions and its intensification of the arms race. It was none other than the current President who postponed the ratification of the Soviet-American agreement on SALT II. And it was under pressure from Washington that the US's NATO allies decided to produce and install in Western Europe the new American nuclear missiles of intermediate range that are aimed at the USSR." [17]

16. A. Tolkunov, *Pravda*, August 13, 1980.
17. *Pravda*, August 16, 1980.

As for the Republican Party, the Soviet press noted, as might have been expected, that its electoral platform had been "dictated by the military-industrial monopolies and . . . the most reactionary circles of American imperialism." [18] Still, mindful of the fact that what politicians say and write is not a sure guide to their actions, the Soviets refused to be unduly disturbed by the Republicans' tough anti-Soviet rhetoric and their call for substantial increases in the US armory of nuclear and conventional weapons.

In fact, amazing as it may seem, it appears that Reagan was the Soviets' favorite in the presidential race. To be sure, when it comes to foreign politics, the Russians seldom show sympathy to a loser or choose to be offensive toward a winner. Still, in their postmortems on the American elections, one could sense a special delight over Carter's defeat and more than a modest hope that Reagan was a man with whom the USSR could do business. Carter's loss, the Soviet press asserted, was due to his dismal record, broken promises, and general ineptitude. But it was his perfidious foreign policy that had especially repelled the voters. One correspondent quoted approvingly an American analysis of the losing candidate's performance in the foreign field: "Everything he did was out of place. He would suddenly bolt from one position to another, and engage in sabre-rattling. He timed Presidential Directive No. 59 about 'limited nuclear war' so it would be publicized on the eve of the Democratic Convention. As a result [most people] stopped trusting him." [19]

Reagan, the same article admits, had indulged in some intemperate language about international relations. But that had been probably just "campaign oratory." And as the race wound up, the President-elect's tone had grown moderate and constructive. "Though Ronald Reagan . . . has not presented a concrete program, his statements have made a very positive impression, and judging by the reaction of public opinion, brought him fresh support from the voters. [Reagan said] . . . that as a man who had lived through several wars, he abhors

18. *Ibid.*, July 19, 1980.
19. "The USA: What the Elections Have Shown," *Pravda*, November 7, 1980.

[the thought of] nuclear catastrophe, and that he is 'for negotiating with the Russians.' "

The writer was also reassured by the people reputedly close
to the future President. "In Ronald Reagan's entourage are
found such experienced public figures as former treasury secretary William Simon, economists Alan Greenspan and Milton
Friedman, and other well-known people. Among them is also
former secretary of state Henry Kissinger, for whom some local observers prophesy a new political career." Well, the Soviet journalist was not as well informed as he might have been
about American politics. But it is interesting to see the hint
that, much as they had crossed swords with him, the Soviets
would not have been adverse to Henry Kissinger's returning
to the leadership of American foreign policy. And it is amusing to note their evident high esteem for Milton Friedman, the
high priest of the most conservative brand of capitalist economics—testimony once again of how little ideological considerations weigh with the Russians in their appraisal of
Western statesmen and men of affairs, and how even the question of whether the individual is "hard" or "soft" on the Soviets is deemed by them secondary to whether he will be consistent and predictable in his approach.[20]

The year 1981 began with Soviet troops still embattled in
Afghanistan and the Polish turbulence still far from subsiding.
A major international crisis was alleviated, but this did not
touch directly on the USSR: Iran finally released the American
hostages. Moscow's reactions had to be mixed: on the one
hand, America's discomfiture had been a source of considerable satisfaction; on the other, a prolongation of the hostage
crisis into the new presidency might well have led to vigorous
US measures against Iran and thus threatened a Soviet involvement. And so there was both regret and relief in the Soviet press account of the release of the American prisoners, as
well as a final salvo of abuse directed at the now ex-President,
Carter. It was he, the Soviets once more insisted, who had pro-

20. Almost needless to say, a *Pravda* article or dispatch on foreign affairs
does not express merely its writer's opinions but must be in line with the
official viewpoint.

voked the entire hostage crisis and then tried shamelessly to exploit it for his personal political ends. "And so the problem of the hostages has been resolved. But one cannot say the same . . . about the US-Iran crisis, the deeper causes of which are rooted in America's imperialist policies." [21]

Whatever the underlying character of US policies, the Kremlin was ready to give the new administration the benefit of the doubt. "In his [inaugural] speech, Ronald Reagan did not state how the new administration intends to conduct itself vis-à-vis the Soviet Union. Our country's foreign policy is based on the conviction that all states must strenuously strive to bring about a healthier international environment and to strengthen peace. It is self-evident that these goals would be served by a positive development in relations between the US and USSR, and their constructive cooperation in resolving the pressing problems of international life." [22]

Perhaps the first two or three months following Reagan's inauguration *could* have been a propitious time for holding exploratory US-Soviet talks. Moscow was facing troubles on several fronts: the Afghanistan venture was proving more difficult than had been expected; the situation in Poland was ever more alarming and the Soviet economic picture and prospects far from cheerful. And, in the background, lay the still intractable dilemma of China. As for the United States, it had just shed the debilitating burden of the hostage crisis, and Reagan's decisive victory as well as his party's having secured a majority in the Senate had gone far to convince the Soviets that the period of weakness and disarray of the executive branch of the American government that had started with Watergate was finally over.

There were thus cogent reasons for the Kremlin to adopt a more accommodating position towards the US and try to conciliate the new administration before its harsh rhetoric about the USSR could crystallize into actual policies and arms programs. Of course Moscow would not supplicate for negotiation or volunteer major concessions—that would have been an

21. A. Tolkunov, *Pravda*, January 22, 1981.

22. "Review of the International Situation," *Pravda*, January 25, 1981.

avowal of weakness and anxiety not in the style of Soviet diplomacy. But an exploration of the Russians' attitudes might well have shown them to be flexible on a range of issues such as revising SALT II, the balance of forces in Europe, approaches towards a settlement in the Middle East, etc. At the moment America's bargaining position as compared with that of the USSR appeared quite favorable.

But while some of his predecessors had shown themselves upon their inauguration overeager to talk immediately with the Russians, Reagan was reluctant to try to initiate a fresh dialogue. Part of the difficulty lay in the cumbrous process of putting together a new administration; its chief aides, especially in foreign affairs, were preoccupied with defining and defending their respective spheres of competence. But the reluctance to approach Moscow was also grounded in the general philosophy of the President's entourage. Many of its members had come from the Committee on the Present Danger, which had opposed SALT II and criticized the whole course of American policy towards the USSR ever since the inception of detente. These people, now in positions of power or influence, tended to feel that any talks or negotiations would be barren of results until and unless the US strengthened substantially its military potential vis-à-vis the USSR.

Yet even if the exploratory talks had proved fruitless, the mere fact of holding them would have had a positive effect on America's European partners, always anxious that their trans-Atlantic ally should in its dealings with the Soviet colossus preserve the proper balance of firmness and conciliatory spirit. Once critical of Carter's alleged indecision and naiveté in coping with Soviet policies, the Europeans were soon complaining that the new administration's approach was too rigid and uncompromising. Well, any course, even the most sophisticated, adopted by the United States still would have encountered criticism on one count or another. One cannot help reflecting, however, that in March or April 1981, America's bargaining position could have been stronger than it was, or at least appeared to be, several months later. In the meantime the Soviets had gone on adding to their arms, while continu-

ing to talk peace and detente. By contrast Washington kept up
its harsh rhetoric about the USSR, but its projected new arms
systems stayed on the drawing board, their rationale and their
enormous costs running into mounting criticisms at home and
abroad. There is a saying that one arms in order to be able to
negotiate. It would but slowly dawn on the Reagan Administration that under the circumstances the reverse was also true.
By themselves neither arms nor agreements can in this nuclear age provide mankind with a sense of security; it is only
a combination of both that can assuage his fear of war and
convince him that his government is acting responsibly.

But whatever advantages accrued to Moscow because of US
reluctance to negotiate, the Soviet Union's image as a patient
seeker after peace and detente could not be convincingly sustained in view of its attitude towards the Polish crisis. While
repeatedly abjuring any intention of intervening in the internal affairs of Poland, the Soviets in fact continued to wage a
relentless war of nerves upon its people.

"What is Solidarity up to in fact?" This question, according
to the Soviet press, was being asked by an increasing number
of Poles, concerned allegedly over the union's disregard of its
original pledges and its growing interference in purely political matters.[23] And, in the view of a commentator, it was becoming clear that at least some within Solidarity "are trying
to overthrow the socialist system in Poland and take the country out of the socialist alliance of the Warsaw Pact Organization. . . . Under the mask of socialist 'renewal,' [foreign and
domestic] counterrevolutionary forces are striving towards the
ultimate goal of destruction of socialism in the country."

The comment was typical of that mixture of fiction and half-
truths that characterized the Soviets' intepretation of Polish
events from the beginning of the year until the military coup
of December 13. Needless to say, Solidarity could not avoid
political involvement: no major social or economic questions
in a Communist country can be separated from politics. But
no one in Solidarity, no Pole in his right mind, could think of

23. "About the Developments in Poland," *Pravda*, January 30, 1981.

tampering with the socialist framework of society, or of turning state-owned industry and banking over to nonexistent
capitalists, or of re-creating large landed estates, most of whose
former proprietors had long since died in foreign parts. Admittedly the Polish people as a whole had never been overly
enthusiastic about their country's ties with the Soviet Union.
But a great majority of them had come to accept the ineluctable facts of geography and history: Poland had to stay within
the Warsaw Pact. From the beginning the leadership of Solidarity eschewed even a hint of anti-Russian rhetoric or of
questioning Poland's membership in the Communist alliance.
As the year progressed it became evident that the same caution was not being observed by the more hot-headed among
Solidarity's rank-and-file members. Yet it is fair to say that if
allowed to preserve a degree of internal freedom, most Poles
would have acquiesced in the continuation of the Communist
regime and in the country's remaining within the Soviet bloc.

But it is unlikely that any diehard Communist, let alone one
in the Soviet Politburo, would have been impressed by such
considerations. He would find it incredible that a *real* Communist regime could long survive amidst growing political
anarchy and license, and it was thus that the current Polish
scene was viewed from the Kremlin. Nor could the latter tolerate even the mere possibility of the military link between
Poland and the USSR—the very core of the Warsaw Pact Organization—being dependent on the free choice of the Polish
people. What was happening in Poland between August 1980
and December 1981 must have been so repugnant to the Soviet leaders that it is a tribute to their self-control and powers
of simulation that they managed to preserve an outward facade of composure and relative unconcern, hoping against
hope that the events in that always troublesome country would
sort themselves out and that the Polish comrades would eventually regain control over the situation.

Much as Moscow considered military intervention as a
weapon of last resort, it still could not explicitly disavow the
possibility. And even though veiled in double-talk, this threat
in turn made even the more faint-hearted among the NATO
governments realize the need of strengthening Western Eu-

rope's defenses, including installing intermediate-range nuclear weapons there. The Polish crisis added plausibility to the American warnings about the Soviets' intentions. Already in the waning days of the Carter Administration, a White House statement had stressed that "preparations for possible Soviet intervention in Poland appear to have been completed."[24] And very soon the Kremlin had an occasion for some second thoughts about its initial satisfaction over Reagan's election: "In connection with the USSR, Ronald Reagan has allowed himself some expressions more venomous in their tone than those heard from any previous head of the Washington administration,"[25] noted a Soviet observer on January 31. Indeed, the Soviets' discomfiture over Poland offered the US similar opportunities to those that the Iran hostage crisis had placed before the Russians: the opponent could be accused of heinous imperialist intentions and designs, while at the same time be derided for his indecision or his inability to put them into effect. But the irony of this reversal of situations was not being appreciated in Moscow. Unlike those of the Czechoslovak crisis of 1968, the ramifications of the Polish one transcended the boundaries of Eastern Europe and the Soviet bloc. The crisis was damaging the Soviet Union's overall political and diplomatic interests. The threat of intervention was weakening the credibility of the Soviets' professions of peace, while the fact that as yet it had failed to materialize tended to weaken the other side of the USSR's image: that of a fearsome military colossus that could not be challenged with impunity.

As time went on fear of the Soviets was also bound to become less effective as a brake on the Polish revolution. The workers' demands escalated, and clashes with the regime continued as it tried in many cases to go back on the promised reforms or temporize about putting them into effect. But the main cause of the continuing strife was the country's steadily worsening economic situation. In trying to cope with growing inflation caused both by its mismanagement of the economy and by the concessions it had been recently forced to grant,

24. *New York Times*, December 8, 1980.
25. *Pravda*, February 1, 1981.

the government had to resort to drastic price increases. These in turn were bound to lead to new protests and strikes. Industrial and agricultural production kept falling. As the winter of 1980–81 ground on, the most typical Polish urban scene became long queues in front of meagerly stocked stores.

The nation's initial feeling of elation at the victory over the regime and the freer political atmosphere was now tempered by equally widespread exasperation about the material conditions of life. Had the government scrupulously kept all its promises, which was far from being the case, it still could not have performed the economic miracle needed to appease the people's frustrations. Thus the strife continued, both within the party, where the reformers clashed with those who would have turned back the clock, and between the regime and the society at large. In both camps there were people who urged that only a sensible compromise between what the nation wanted and what was practical in view of Poland's situation could save the gains of the revolution and minimize the danger of an invasion. Yet there were equally persuasive voices arguing that no matter what stand Solidarity took and what the character of the reforms, the Russians would march in if and when they felt they could afford it, and that one invited defeat by trying to shape one's policies while worrying what the Kremlin might or might not do. Conversely, to a small group of Communist hard-liners, the prospect of an overt or covert intervention by the Soviets appeared to be the only way out of their present predicament. Hence they echoed the Soviet charges that Solidarity had been infiltrated by agents of foreign imperialism and that it was anti-Russian insofar as its ultimate goals were concerned.

In February Gen. Wojciech Jaruzelski was appointed prime minister, while retaining his post as minister of national defense. The appointment was received rather favorably by public opinion. In Polish history a military uniform has always carried nationalist connotations, and there were rumors that Jaruzelski had declared the previous August that he could never order his soldiers to fire at the striking workers. There was, on the other hand, the fact that the general had for a long

time been a member of the Communist party and of its Politburo and, as was the case with every high-ranking officer of
the Warsaw Pact states, his military advancement would not
have been possible without Moscow's express approval. And
so his appointment might well have been made for other reasons than just placating Polish public opinion.

A few days after his elevation Jaruzelski, along with Kania,
attended the Twenty-sixth Congress of the Communist Party
of the USSR. At the Congress they heard Brezhnev in his General Secretary's report speak sternly and ominously about the
Polish situation: "In fraternal Poland enemies of socialism,
helped by foreign forces, have been instigating anarchy, thus
trying to turn the course of events in a counterrevolutionary
direction. . . . [But] the Polish Communists and the working
class can fully rely on their friends and allies; we shall not
abandon fraternal socialist Poland in its plight. We shall not
allow it to be ruined. . . . Communists have never faltered
before the enemy, and they have always prevailed. Let no one
doubt our firm resolve to protect our interests and safeguard
the socialist achievements of our nations." [26]

Even though Brezhnev's ringing warning was greeted by ap-
plause, some at least among the thousands of delegates (certainly among the foreign Communist guests) must have privately felt that the speaker had gotten his facts all mixed up.
How could socialism in Poland be in jeopardy when what one
was witnessing there was (except for its nonviolent character)
a classical example of the Marxist model of revolution: virtually the entire working class rising against the exploiters—
the party oligarchs and bureaucrats? And how about those
sinister outside forces abetting anarchy and counterrevolu-
tion? The more sophisticated in the audience knew that the
capitalists had lent the Warsaw government more than $25
billion, and it was not their fault that these enormous sums
had been squandered by the Communist bosses. In fact the
wish that law and order of the pre–August 1980 variety return
to Poland was probably as keenly felt in certain Western bank-

26. *Twenty-sixth Congress of the Communist Party of the Soviet Union*
[stenographic report] (Moscow, 1981), vol. 1, p. 26.

ing circles as it was within the Soviet hierarchy. And while
the struggle of the Polish people aroused world-wide sympathy, in Western Europe this sympathy was tempered by apprehension lest the Kremlin in its wrath be driven to some desperate measures, perhaps not only against Poland.

It was not only the Polish crisis that cast a shadow over the
congress. In his speech Brezhnev had to acknowledge much
more explicitly than on any previous occasion the existence
of serious dissension and of criticism of the USSR within the
international Communist movement. Moscow, he protested,
was far from wishing to dictate to the fraternal parties: "Nobody wants to impose on others rigid schemes and programs
that would ignore the individual characteristics of this or that
country." [27] But this apparent concession was immediately
qualified: there could be no compromises when it came "to
the conflict of principles between revolutionaries and reformers, between creative Marxism and dogmatic sectarianism."
And obviously it had to be the Kremlin that decided which
was which.

The same approach was followed in the General Secretary's
discussion of the foreign Communists' criticisms of the Soviet
system: "We do not assume that everything in our country is
perfect. . . . We listen attentively to comradely, constructive
criticisms. But we resolutely reject the kind of 'criticism' that
distorts the socialist reality [in the USSR] and thereby, whether
intentionally or not, serves the purposes of imperialist propaganda and of the class enemy." [28] It could be no secret to whom
this admonition was addressed. In listing the nonruling Communist parties with which the Soviet one had maintained good
relations, and that consequently deserved high marks for good
behavior, Brezhnev ostentatiously omitted those of Italy and
Spain. To a veteran of the Stalin era, the whole discussion
must have appeared unreal: how could there be such a thing
as a Communist party that did not obey, let alone criticized,
Moscow? One had grown unhappily accustomed to the insol-
ence of the Chinese. But it was a relatively novel and unset-

27. *Ibid.*, p. 35.
28. *Ibid.*, p. 34.

tling experience to have alleged Italian and Spanish comrades
indulge in brazen attacks upon the Fatherland of Socialism,
castigating its leadership for what it had done in Afghanistan,
threatened to do in Poland, and was actually doing to the dissidents among its own people. And some among the younger
delegates may have reflected that, far from being a source of
strength and support to their regime, international Communism was turning into more and more of a burden.

If somewhat dispirited by Brezhnev's survey of the Communist world, participants in the congress were probably
heartened by what he had to say about the situation in the
capitalist one. In fact one could not question his analysis of
the current dilemma of the Western economies: "Measures
taken by the bourgeois governments in order to fight inflation
contribute to the stagnation of production and the growth of
unemployment; attempts to stop the fall in production tend to
intensify inflation."[29] Equally satisfying to a true believer
should have been another economic item reported by Brezhnev: over the past ten years, the US share of the world market
had declined by almost 20 percent, as Japanese and Western
European exports increasingly challenged America's industrial supremacy. According to the canon of Marxism-Leninism, economic crises strengthened imperialist tendencies
within the capitalist world. Therefore it was not surprising
that the bourgeois states had been spending more and more
on armaments. "In the US such expenditures have reached
$150 billion, and even this astronomic sum is not enough for
the American military-industrial complex. It clamors for
more."[30]

As he continued his survey of the international situation,
the General Secretary switched from themes designed mainly
for domestic consumption to others addressed to the outside
world. How foolish and dangerous for some American strate-
gists to believe that limited nuclear war was possible: "Such
a 'limited' war . . . say in Europe, would mean in its very
beginning the absolute end of European civilization. And it is

29. *Ibid.*, p. 37.
30. *Ibid.*, p. 38.

obvious that the United States would not be spared by the flames of war." [31]

Brezhnev was even more emphatic in his attempt to discredit another thesis, allegedly espoused by the new Washington administration, whose main argument was that having achieved strategic superiority, the USSR believed it could meaningfully win a full-scale nuclear war. "What is this Soviet military superiority about which they talk? The danger of war hangs over the United States just as it confronts every other state in the world. But the source of that danger is not the Soviet Union and its mythical superiority, but the arms race itself. . . . We are ready to struggle against this real danger, combat it arm in arm with America, with the European states, and with all other countries on our planet. But it would be sheer madness to try to outdo one another in an arms race, or to count on achieving a victory in a nuclear war." [32]

The Russian leader was equally skillful in parrying other allegations concerning the Soviet Union's military and expansionist goals. No, the Kremlin did not try to withhold from the Soviet people knowledge of how horrible a nuclear war would be: "We propose the formation of an authoritative international committee that would discuss the desperate need to avoid a nuclear catastrophe. . . . The committee's conclusions should be spread throughout the world." [33] Then, he continued, what were all those slanders about the Soviets planning to grab control of the Persian Gulf? On the contrary, the USSR stood ready to conclude a treaty "that would guarantee the sovereignty of all states in the region, as well as the security of sea and other channels of communication linking the area with the rest of the world." [34] And to put at rest West European fears, unreasonable as they were, "We propose an immediate moratorium on the deployment by NATO and the

31. *Ibid.*
32. *Ibid.*, p. 40.
33. *Ibid.*, p. 48.
34. *Ibid.*, p. 39.

USSR of new missiles of intermediate range; that is, a freeze both quantitative and qualitative on such weapons."[35]

To be sure, when one took into account the number of qualifications Brezhnev attached to his proposals, they had to appear less than completely reassuring. Thus he would let Afghanistan be placed on the agenda for the discussion of the security of the Persian Gulf. But it was only Afghanistan's foreign relations that could be discussed, not its internal affairs. To touch on the latter (i.e., the Soviet military occupation of the country) would be incompatible with "the sovereignty of Afghanistan and its status as a non-aligned state."[36] Under the prospective moratorium on intermediate-range nuclear missiles in Europe, American air- and seaborne weapons in the area would have to be included in the NATO quota, but on the Soviet side, Brezhnev implied, only land-based missiles would be "frozen."

Such escape clauses notwithstanding, Brezhnev's "peace initiatives" added up to an effective performance when compared with the Reagan Administration's indecision as to whether to talk with the Russians and what to discuss. (Indeed in April the Reagan Administration would lift the grain embargo imposed by Carter after the invasion of Afghanistan, but it would still balk at comprehensive negotiations with the USSR.) The "ifs" and "buts" of the Soviets' proposal could be detected only by the initiated, and the ordeal of trying to wrest a meaningful concession from them at a conference table could be appreciated only by those who had had that experience. To the world at large, it may well have appeared that Brezhnev was proposing concrete steps to relieve international tension, while the State Department's by now almost ritualistic brusque rejoinder—there was nothing new in the Soviets' proposals—seemed a sign of bad faith or of lack of self-assurance on Washington's part.

Yet even if Brezhnev's pleas for negotiation were seen as a propaganda manoeuvre (which they largely but not entirely

35. *Ibid.,* p. 47.
36. *Ibid.,* p. 46.

were), there were still cogent reasons for the US to come forth with its own counterproposals. The crisis within the Soviet bloc offered a good opportunity for an imaginative American initiative: should not East and West confer about possible joint measures to help Poland out of its growing economic distress, with the West denying any intention of seducing the country away from the Warsaw Pact and the Soviets pledging that they would not invade?

Such a move would have put Moscow on the spot. The USSR could have denounced the American proposals as a provocation and repeated the injunction that the capitalists had no right to interfere in the internal affairs of a socialist country. But in terms of propaganda, the offer of Western help would have made it much harder for the Soviets to keep up the fiction that sinister capitalist forces were seeking Poland's ruin. And there was just an outside chance that in their current indecision about what to do about the crisis, the USSR and the Polish regime would have accepted the West's offer, with results that could not but be beneficial for everyone concerned, as well as for European peace and tranquility. Instead Washington and other NATO powers continued to wait upon events, while persisting in warning the USSR that invasion would be attended with dire results for every aspect of West-East relations.

The Kremlin, as we have seen, did not need Western sermonizing to appreciate the risks and costs of an invasion. The Soviet-preferred solution to the Polish crisis still remained an internal one, i.e., without *overt* Soviet participation. Superficially, as the year progressed, chances for such a solution seemed to grow dimmer. The Warsaw regime was being confronted by new demands, and in many cases forced to promise fresh concessions. The workers clamored for, and in principle were eventually granted, a voice in the management of industrial enterprises. Farmers came up with a union of their own which the government, after fighting a delaying action, finally had to approve. There was steady pressure on the authorities to dismiss those provincial officials and industrial managers who resisted the reforms or who retained attitudes that in the eyes of the workers were incompatible with the spirit of the times.

As against this general state of things—which, in the eyes of an old-fashioned Communist bureaucrat, looked like a veritable Sodom and Gomorrah—there could be perceived signs that the regime was readying contingency plans for a possible counteroffensive. Upon becoming prime minister, General Jaruzelski began to place military personnel in key positions within the civil administration. The Soviet commander-in-chief of the Warsaw Pact Organization, Marshal Victor Kulikov, became a frequent visitor to Poland, conferring with high party and government officials. The spring 1981 manoeuvres of the Pact's forces lasted unusually long and extended over parts of Poland as well as East Germany and Czechoslovakia.

By and large, such portents were being ignored by the population. The leadership of Solidarity had its hands full trying to consolidate the gains under the reforms and to restrain the more impatient among its members. The average Pole was too preoccupied with the struggle to secure the daily necessities of life to pay much attention to this apparently aimless reshuffling of ministers and officials. After its performance of the last few months, few could credit the government with the will, let alone the ability, to reverse the country's course. And so fears still centered on the possibility of a Soviet move. But the Warsaw Pact manoeuvres came and went, and the "fraternal" armies' troops returned to their own countries.

Even though its patience was being sorely tried, the Kremlin had no cause for undue haste. For the moment and the foreseeable future, there was little danger of the Polish disease spreading to the other countries of the Soviet bloc. The Czech or Hungarian worker might indeed aspire to the rights and freedoms being won by his Polish counterpart, but he would hardly try to emulate his example as long as Poland's economic situation went from bad to worse. To be sure the country's economic plight cut both ways. The USSR could not view with equanimity the prospect of Poland's industrial production grinding to a halt and of its standard of living declining to a point where the people would forget their inhibitions against violence. Help from the USSR as well as other Communist countries had to be rushed in. According to a Communist source, between September 1980 and May 1981 the USSR alone provided Poland with goods and hard currency amounting to $4 or $5 billion.[37]

Another of Moscow's immediate concerns had to be to minimize the damage being done to the Polish Communist party itself. Buffeted by the forces of reform and nationalism, the party had not only suffered losses in authority and membership, but also stood in danger of being transformed into something quite different from the traditional rigidly hierarchical Marxist-Leninist model. An extraordinary party congress was to assemble in July. But the manner of electing delegates to it departed radically from the traditional Communist pattern. Instead of lists of nominees being prepared by the central organs and then "elected" unanimously by the local membership, there were to be genuinely free elections! Throughout the spring of 1981 the rest of the Communist world watched with amazement bordering on horror, as the electoral meetings often turned into tempestuous debates, with party higher-ups being berated by the rank and file, with many a central committee bigwig being denied a mandate to the congress. If this was allowed to go on the congress was bound to resemble a Western party convention, and the new leadership (i.e., the Central Committee and the Politburo) it would elect might turn out to be hardly different in its attitudes from Solidarity.

Occasionally local electoral meetings heard speeches touching on subjects hitherto considered taboo. It was startling to hear a Communist state quite explicitly that it was Soviet exploitation that had been largely responsible for Poland's economic plight. "One thing is certain: the party must find the courage and strength to eliminate the causes of the present crisis, [the chief of which] lies in the unequal terms of trade [with the USSR], terms that border on open robbery." The party, the speaker also insisted, must thoroughly heal itself,

37. *Trybuna Ludu* ("The People's Tribune," the official Communist party organ published in Warsaw), June 12, 1981.

for otherwise it would be reduced to "a body of a few hundred thousands, most of them administrative, security, and military officials—that is, the people who feel revulsion against physical labor." But it was not any pro-Western or antisocialist sentiments that inspired this outburst, for this unusually candid Communist denounced also "the class enemy and enemies of the nation" in the West "who would like to see Poland become another Afghanistan." [38]

On June 5 the Central Committee of the Soviet Communist Party addressed a letter to the Polish leaders, urging them to take decisive steps to curb "the enemies of socialist Poland" and to prevent the coming congress from presiding over the death of Polish Communism. "A wave of anti-Communism and anti-Sovietism is gaining force. . . . The party can and ought to change the course of events even before the Ninth Congress." [39] The letter, made public in Poland a few days later, did have some of its intended effect. The regime stiffened its stand and exerted whatever influence and power it could summon to secure a more orderly (from the orthodox point of view) electoral process. At the congress, those who wanted to transform drastically Polish Communism found themselves in the minority. The Kania-Jaruzelski leadership was confirmed, and while some of the most notorious hard-liners bit the dust, the majority of the new Politburo was still cast in the pre–August 1980 mold. For the moment the Soviet leaders could breathe more easily. They had probably never been so close to deciding to invade as in May and June when the Polish party, the main instrument of their past—and hopefully future—control of the country, was on the point of getting completely out of hand.

While the worst had been avoided, the Kremlin still had to feel a sense of urgency about the situation. The impasse between the regime and the society had not been resolved. The government had ceased being feared, but it had not succeeded in convincing the people that it ought to be trusted. Solidarity

38. Quoted in *Kultura* ("Culture," a Polish-language periodical published in Paris), November 1981, p. 158.

39. Quoted in *Trybuna Ludu*, June 11, 1981.

had been carrying out a revolution, but it had neither the will nor the means either to stop the revolutionary process or to consummate it by taking power. The Soviet Union remained unwilling to intervene, but it felt it could not repudiate unequivocally the threat of intervention.

In their mutual frustration, both contending sides in Poland escalated their rhetoric. The regime now invoked more and more frequently the spectre of "national catastrophe" (read Russian tanks) should Solidarity persist in its demands. The latter, in turn, was proving less amenable to counsels of caution by Walesa and other moderate leaders. The free union's national congress, held in September, issued a message "to the working people of Eastern Europe" urging them to follow the Polish workers' example.[40] Solidarity also demanded "social control of the mass media," as well as several other things that, although at face value unsensational, were incompatible if not with the theory then with the practice of Communism.

It is unlikely that such imprudent declarations, or the even more fiery sentiments expressed during the Solidarity Council's meeting in December, were the direct cause of the subsequent proclamation of martial law. Plans for it must have been drawn up long before, and the actual decision probably reached after the July party congress.

On October 18 Jaruzelski added to his panoply of offices that of the First Secretary of the Communist party. For a while the regime continued to mask its intentions, as the First Secretary–Prime Minister sketched vistas of a broadly based national front that would include the Communist party, the Church, and Solidarity. Then, on December 13, this seemingly indecisive and ineffective regime struck. Martial law was proclaimed throughout Poland; thousands of Solidarity activists, as well as dissidents from every walk of life, were seized and imprisoned. During the state of emergency, Poland was to be ruled by a supreme military council headed by Jaruzelski. Political agitation and incitement to or participation in strikes were made subject to drastic penalties to be meted out by courts martial.

40. *Ibid.*, September 10, 1981.

"TASS has been empowered to declare that the Soviet leadership and people are following attentively the Polish developments. . . . We greet with satisfaction the statement of W. Jaruzelski that the Polish-Soviet alliance has been and will remain the foundation of the Polish national interest and the guarantee of the inviolability of the country's frontiers, and that Poland will remain an inseparable member of the Warsaw Pact and of the socialist commonwealth of nations."[41] This official and emphatic approval of the coup leaves little doubt that the USSR had foreknowledge of what was going to happen, and indeed it must have assisted with the plans for the military takeover. It is equally likely that the Soviets had stood ready to back the coup with their own forces in case it met with armed resistance on the part of the population or led to mutinies within the Polish army.

But no such resistance in fact materialized. Solidarity as well as the nation as a whole was taken by surprise, and once the initial disbelief passed, it gave way to sullen resignation. So at least it appeared during the first months of martial law. One element in the numbing stupefaction that at first seized the people was the docility with which the Polish soldier let himself be used to stand guard over his countrymen. To be sure, it was the special security forces who were used for police action, with the regular army units being held in reserve. And the story might well have been different had the soldiers actually been ordered to fire upon the workers. But here the flamboyant rhetoric of some Solidarity leaders in the weeks preceding the clampdown did harm to the nation's cause. To repeat, this rhetoric could hardly have influenced the decision to institute martial law: the meticulous attention to detail and the efficiency with which the coup was executed show clearly that it must have been planned months in advance. But the reckless talk by the regime's critics gave at least a shadow of plausibility to the patriotic rationale provided by Jaruzelski and Co. for their action: that the Polish army had to take over because otherwise the Russians would surely have moved in.

The ease with which the coup was carried out and the rel-

41. *Pravda*, December 15, 1981.

ative absence of active resistance during the first few months
of military rule were, however, deceptive. Polish society had
been taken into custody, but neither had it been pacified, nor
had it acquiesced in the loss of the rights and freedoms it won
between August 1980 and December 1981. The military regime, or more properly Polish Communism in uniform, was
unable to win over any major component of society, or to convince the people that it was much more than a surrogate for
Soviet occupation. The Polish economy was far from showing
any signs of recovery. The Church as before preached to the
people the need for patience and nonviolence and urged the
rulers to seek national reconciliation. But as of the summer of
1982 the chasm between the regime and society remained
much greater than it had been before December 13. And unhappily it was easier to foresee this confrontation continuing
and even assuming violent forms than to anticipate the chasm
being bridged by a meaningful compromise.

Moscow, as well as the Warsaw regime, had then to view
the coup and its sequel as a palliative, rather than a long-term
solution to the Polish problem. And for the USSR the broader
implications of the problem posed some very basic questions
touching on the overall character and direction of Soviet foreign policy.

In the short run the Kremlin's expectations that indirect
aggression would cost much less than a full-scale invasion
were amply justified. The coup was generally condemned
throughout the world, but the condemnation was combined
with relief that, for the present at least, it was not *Soviet* tanks
that were patrolling the streets of Polish cities. At first almost
everyone agreed that detente had suffered a fresh blow, but
very soon voices, notably in West Germany, were urging that
issues such as SALT and the possibility of a nuclear moratorium between NATO and the Warsaw Pact ought not to be
affected by what the Soviets were or were not doing in Eastern
Europe. Even one as sensitive to the Soviet threat to Western
Europe as Helmut Schmidt joined those arguing that the die
had been cast at Yalta in 1945, when Roosevelt and Churchill
allegedly agreed that Eastern Europe should become the pre-

serve of the USSR and Communism.[42] While NATO censured
the actions of the Jaruzelski regime, both West Germany and
France reasserted their intentions to continue their joint commercial ventures with the USSR. In the face of strong American objections, the two governments thus proposed to go ahead
with plans to help construct and finance the giant gas pipeline. Proposals for stopping the flow of fresh credits to Poland
and for calling in the old debts, thus in effect placing on the
USSR the burden of saving the country from bankruptcy, were
met by warnings that Western economic pressure might drive
the Warsaw regime into the Russians' arms! In brief there was
the usual (as of late) reluctance in the West to recognize the
facts as they were. And underlying at least some of these rationalizations was the quite evident fear of unduly offending
and provoking Moscow.

But it was not the immediate consequences of the crisis
within its empire that had to be of the greatest concern to the
Kremlin as it faced the prospects for the eighties and beyond.
Poland, whether in a state of unrest or submission, simply
epitomized the intractable dilemma that challenged the entire
rationale not only of Soviet foreign policy, but also of the whole
Soviet-Communist world system. The dilemma bears some re-
semblance to that of the sorcerer's apprentice who, having
conjured up a destructive force, found himself unable to devise a formula that would make it subside or obey his com-
mands.

One such force summoned up by the Soviet Union in the
post–World War II era, as it tried first to counter the predominant position of the West and then get the upper hand, was
the power of nationalism and anti-imperialism. The process of
Western retreat from imperial power would have been difficult and marred by violence in any case, but Soviet policies
and propaganda were intended to make it even more so, while

42. This at best is a gross oversimplification. The Big Three proclaimed at
Yalta that the liberated nations should be free to determine their own forms
of government, and if there was an implication that Poland, Czechoslovakia,
etc., were to be comprised within the Soviet sphere of influence, it was still
assumed that there would be no constraints on their internal freedom.

behind the facade of anti-imperialism and the sponsorship of wars of national liberation, the USSR tried to consolidate and expand its own empire. By the seventies one could see quite clearly that Moscow's efforts were not designed merely to destroy what remained of the old world order, but to prevent any system of international stability from coming into being. There was to be no detente, if the USSR could help it, between the West and the Third World, even though the Kremlin sought to reach accommodation with the West itself.

Yet by that time it was even more obvious that Communism itself was far from being immune to nationalism and anti-imperialism. China's estrangement and defiance could no longer be represented by Moscow as a temporary aberration wrought by the machinations of the malevolent "Mao clique." It had to be seen as a manifestation of an organic ailment of the world Communist system, one which could not be cured by ideological incantations.

When it came to the smaller Communist states and Moscow's Third World clients, the Kremlin still continued to believe that Russia's military power was an effective remedy against outbreaks of local nationalism and anti-Sovietism. This had certainly proved true in Hungary in 1956 and Czechoslovakia twelve years later. But armed force could not be the answer to all of the intra-bloc troubles.

A reflective Soviet reader would not have missed the fact that the great festival of Soviet imperialism on the occasion of Leonid Brezhnev's seventy-fifth birthday in December 1981 was marred by some conspicuous absences. Missing from the roster of leaders of the fraternal countries who flocked to Moscow to lavish eulogies, orders, and presents upon Leonid Ilyich were the current heads of the Albanian, Yugoslav, and Polish regimes. The first persisted in its heroically preposterous stance of equal hostility towards Russia and China, not to mention the capitalist world; Albania was probably the only country in the world to deserve the promiscuously used term "nonaligned." Yugoslavia, though no longer hostile, had spurned Moscow's enticements to rejoin the Soviet bloc. And the leaders of People's Poland were currently too busy trying to keep its people at bay to join in the festivities.

As if to counter any impression that what had been between 1945 and 1948 a solid and tightly disciplined bloc of satellites was now shrinking and vulnerable, the Soviet leaders chose to advertise their most recent expansionist venture. Television as well as the front pages of the newspapers featured a photograph of Babrak Karmal handing Brezhnev the highest order of Afghanistan, that of the Sun of Freedom. (An imperial system is compounded of many elements, sometimes including both the macabre and the ridiculous.)

But the militant nationalism and anticolonialism that the Soviets abetted against the West eventually were bound to turn against them. To be fair, one must concede that in the post-Stalin era, the Kremlin had tried to consolidate and extend its empire by means other than just military power and ideology. Within both the USSR itself and the allied Communist states, Moscow tried to combine authoritarian politics with a degree of social and economic liberalization and modernization. The socialist bloc was to be held together, and Soviet primacy preserved, by a community of interests and by a rising standard of living in Eastern European nations. To win over Third World countries, the Soviets began to export Russian industrial experts and techniques of rapid industrialization, as well as weapons and anti-Western propaganda. It was by making Communism stand also for a consumer society that the Kremlin sought to prop up its own power as well as to lessen its dependence on political repression and military force. Detente, among other things, involved Moscow's gamble that opening the USSR and Eastern Europe more widely to the West—in fact permitting a degree of Westernization of their societies—would strengthen rather than endanger the political stability and popular appeal of Communist rule.

The Polish experience must have gone far in disabusing the Soviets of such hopes. It had proved difficult for a centralized economy cast in the orthodox Communist mold to put to good use vast foreign credits and expanded trade with the West. In fact both eventually compounded the regime's economic and political troubles.

Equally questionable proved the assumption that economic growth and a rising standard of living would by themselves immunize a Communist society against political turbulence. On the contrary, the "revolution of rising expectations" tended to engender demands for more freedom, rather than inducing political contentment and apathy. And when the rising economic expectations were unfulfilled, the result was something resembling a real revolution.

In brief, neither Russian military power nor economic palliatives have proved fully effective in ensuring the stability of the Soviet bloc. And when one looks into the future, the costs and risks of the Soviet Union standing armed guard over the fragile Communist systems of six European nations with a combined poulation of over 100 million are likely to grow ever greater and more complex. There is an inherent paradox as well as a lesson in contrasting the amount of attention and concern Moscow has to devote to internal developments in Poland or Czechoslovakia with the relative equanimity with which it can view what happens in Finland's domestic politics.

But had the Soviet leaders been capable of absorbing such lessons they would have long ago reexamined the basic premises of their foreign policy, and not only with regard to Eastern Europe. Abroad, Communism in power is no longer a dependable servant or even an asset to the Soviet Union. It may become a threat, as in the case of China, or a heavy burden, as in Poland. And when Soviet expansion is rationalized in terms other than ideological, it brings with it costs and dangers incommensurate with real gains, and even these may turn out to be ephemeral. Witness the case of Egypt. Why then the Kremlin's urge to seek new conquests and clients, if the ultimate consequences of such ventures, rather than serving the interests of the Soviet Union, may well turn out to be detrimental to its power and security?

To the present generation of leaders, this question would in all likelihood appear preposterous. They retain little, if any,

of the sense of ideological mission that inspired their predecessors in the immediate post-Revolution period. Indeed, as
Alexander Solzhenitsyn has suggested, they may be entirely
cynical on the subject. But for practical, if not ideological, reasons, it is impossible for them to abandon the tenet of Marxism-
Leninism that proclaims that recent history must be understood in terms of a struggle between two systems: one epitomized and led by the United States, the other by the Soviet
Union. This struggle does not have to assume violent forms,
and in the nuclear age, it must not, if at all possible, lead to
an all-out war. But to discard the old formula entirely, and to
halt the attempts at destabilizing the capitalist world and expanding the Soviet sphere of domination and influence, would,
in the Kremlin's view, pose a grave danger to the cohesion of
the Soviet system itself. Over and above any considerations of
national security, it is those touching on the preservation of
the present form of Communist rule in Russia that require
Moscow to persist in conceiving of international politics as an
arena of constant struggle, with Communism and its allies advancing and capitalism in retreat. Phenomena that might be
thought to cast doubt on the veracity of this picture—such as
China, or the failure of Communism to sink its roots in Eastern
Europe—are, even though the leaders know better, dismissed
in the official view as but temporary aberrations. Fears that
might be aroused in the Soviet citizen by his leaders' view of
the world are assuaged by official assurances about the "invin-
cible might" of the USSR, as well as by pointing out the growing power of the peace movement in the West. Whatever the
imperialists might be scheming, the "peace-loving masses" of
ordinary citizens in the capitalist countries stand, it is said, as
an additional barrier against the unleashing of a nuclear war.

In brief, the regime believes that its internal security is in-
extricably bound up with the advance of its external power
and authority. For the Soviet elite this advance provides psychological reassurance that for all of its oligarchical, repressive methods of ruling, it still serves the interests of the nation. And for the people at large, Russia's greatness and power
in world affairs are offered as compensation for the failure of

other promises of Communism to come true. With the country's mounting social and economic problems, with the ideology itself having become discredited or irrelevant in the minds
of the great mass of the Soviet people, the regime strives to
demonstrate its viability and dynamism through foreign expansion. It tries, and not without success, to inculcate the lesson that for all of its internal shortcomings and excesses, it has
been under Communism that Russia (and it is at the Russians,
rather than at other ethnic groups in the USSR, that this argument is mainly directed) has steadily advanced in power
and worldwide influence, while the democracies, for all their
alleged freedoms and riches, have been in disorderly retreat,
insofar as their international role is concerned.

Few students of Soviet affairs would consider the picture as
presented above to be immutable. While the general tendency
of Soviet foreign policy is predetermined by the character of
the regime, specific policies are influenced by actual circumstances and personalities. No one should expect Brezhnev or
any conceivable successor of his to declare in effect: "Comrades, I am happy to say that the imperialist danger has disappeared. We can now devote all our efforts to the cultivation
of our socialist garden." But we have seen how reactions, or
lack of them, by Western statesmen have affected Kremlin's
moves on a number of issues, and how the Soviets' calculus
of risks and gains inherent in moving into Angola or Ethiopia
has been influenced by their reactions to Watergate or OPEC's
1973–74 coup. One can speculate how much, but not whether,
Soviet policies towards the US would have been different had
the latter taken up Moscow's hints of a joint front against
China. We have described the dialectic underlying Russia's
thinking about America: the inevitability of rivalry between
the two superpowers is balanced by the recognition that a degree of cooperation between them is necessary so that the most
frightening consequences of that rivalry may be avoided.

It is tempting but unrealistic to postulate that the next generation of Russia's leaders would want or could afford to
change drastically the traditional pattern of Soviet foreign policies and/or seek genuine cooperation with the West—cooper-

ation going beyond efforts to minimize chances of nuclear war and to secure for the USSR the benefits of Western trade and technology. True, whoever is called eventually to replace the present septuagenarian Kremlin team (and the successors' average age may be as low as sixty), they are likely to have a different set of priorities and a different outlook on a number of problems from those venerable relics of the Stalin era, whose main concern has been to avoid its horrors, but otherwise to keep things unchanged. There may be a new Khrushchev among them, willing and daring to shake up the ossified state and party bureaucracy. There will, in all likelihood, be reforms and innovations aimed at curing the perennially ailing segment of the Soviet economy—agriculture—and at stimulating the currently sluggish industrial growth.

Changing the basic pattern of Soviet foreign policy, however, looms as a much more difficult and complex problem. This pattern, as we noted, has become an important element of the rationale of the Soviet system as a whole. The new leaders might well conclude that efforts to propagate Communism in foreign parts have become both anachronistic and counterproductive; that by heightening international tension in areas like the Middle East and Africa the USSR incurs dangers far exceeding any commensurate gains; and that a reduction in the country's nuclear and conventional forces would be both feasible and of great benefit to the economy. But even if they reached such conclusions, our hypothetical Soviet leaders would find it exceedingly hard to act upon them. Applying the brakes so precipitately to the whole process and rhetoric of expansion would throw into turmoil not only the Soviet bloc, but conceivably the USSR itself. Nikita Khrushchev's modest efforts in 1955–56 to liberalize Soviet rule in Eastern Europe had consequences that not only endangered that rule as a whole, but also came close to terminating his own political career. It is unlikely that this lesson has been forgotten by any Soviet politicians.

The formative years of the younger members of the Soviet elite coincided with a period of uninterrupted growth in their country's military strength and world power. One cannot pre-

clude the possibility that some of them, when at the top, would
opt for more rather than less expansionist and militant policies: pressuring the West more vigorously lest it recover from
its present disarray, deciding that military power should be
used uninhibitedly to cure East European Communism of
whatever ails it, and feeling that the USSR has to do something before China becomes a great industrial and military
power. It is easy to imagine a Politburo member arguing that
the USSR could not afford to become less feared abroad and
more liberal internally. It would then be exposed to much
greater pressures both from the outside and at home: growth
of political dissidence; claims for real autonomy, if not independence, for the non-Russian ethnic groups; in fact demands
for reforms incompatible with the survival of the Soviet state
and system. The admittedly serious economic and social problems confronting the Soviet regime do not in themselves provide a clear indication of the path Brezhnev's and his colleagues' successors will take.

But their decisions on foreign policy in all likelihood will
depend on the Kremlin's perception of the condition of the
West. We say condition rather than policies, for, as we have
seen during the past twelve years, it has been Moscow's reading of the strengths and weaknesses, and especially of the degree of cohesion, of the entire community of democratic nations that has been mainly instrumental in shaping the USSR's
foreign policy. To be sure the actual policies of the US and its
allies have been of importance in affecting those of the USSR.
The Kremlin has to react to the size of the American defense
budget; it has to take notice of NATO's proposals and decisions concerning its nuclear and conventional forces; it has a
great stake in the success or failure of US mediation efforts in
the Middle East. The Soviet leaders' attitudes cannot remain
unaffected by an American president's stand on the human
rights issue in the USSR, or by his statement upon taking office that the Russians still aim at world domination and are
confident that they would not only survive, but win a nuclear
war.

The Soviets have learned, however, to discount much of the

American leaders' rhetoric and to pay more attention to the underlying political and economic realities in this country and the rest of the West. It could not have surprised the Kremlin that after beginning his presidency amidst intimations that it was useless to negotiate with the USSR until and unless the US regained its nuclear superiority, Ronald Reagan had to respond to the economic constraints at home and to the urgings of this country's allies, and come up with comprehensive proposals for fresh strategic and other arms limitations talks. And the outcome of any such negotiations will depend much less on the US's stated position at the beginning than on the ability of American diplomacy to match in skill and perseverance that of the USSR, something which, we also have noted, has not always been the case.

But a new nuclear arms treaty, no matter how comprehensive, would, by itself, guarantee neither peace nor world stability. Under any prospective arms reduction agreement both superpowers would still retain the capacity to inflict horrendous destruction upon each other. Arms limitations and summitry cannot basically change the present and increasingly menacing picture of world politics. They can provide the preliminaries but not the essential ingredient of a new and viable international order.

That ingredient can be found only in the West recouping its strength and vitality, as measured by much more than just military criteria. The greatest setbacks suffered within the past twenty years by the democratic community of nations have resulted not so much from any actions by the Soviets as from what the West has done to itself. Western Europe's inability to sustain the momentum towards political integration, the prospects for which had seemed so promising until the early 1960s, has not only led to the present and dangerous military imbalance between NATO and the Soviets; it has had much wider implications—for instance, damaging the prestige and attractiveness of liberal institutions in the Third World. The failure of the great industrial democracies to synchronize their policies both invited and worsened the effects of the OPEC blow. And these and other self-inflicted wounds of the West

have made it much easier for the Soviet leaders to seek answers to their pressing political and other problems not
through domestic and intra-bloc reforms, but in foreign expansion and piling up of arms.

As of today, the practical difficulties standing in the way of
achieving a closer and many-sided cooperation between the
US, the West European states, and Japan are obvious and formidable. But it is even more difficult to expect the Soviets to
change any basic premises of their policies until such cooperation becomes a reality of international life. Military strength
is a necessary ingredient of the prevention of war, but it is
only Western unity and statesmanship that can provide conditions for allaying the East-West conflict and removing the
main obstacles on the road to real peace.

Index

DATE DUE

MAY 2 '89			
GAYLORD			PRINTED IN U.S.A.